LUMINESCENCE CENTERS IN CRYSTALS
TSENTRY LYUMINESTSENTSII V KRISTALLAKH
ЦЕНТРЫ ЛЮМИНЕСЦЕНЦИИ В КРИСТАЛЛАХ

The Lebedev Physics Institute Series

Editors: Academicians D. V. Skobel'tsyn and N. G. Basov

P. N. Lebedev Physics Institute, Academy of Sciences of the USSR

Recent Volumes in this Series

Proceedings (Trudy) of the P. N. Lebedev Physics Institute

Volume 79

LUMINESCENCE CENTERS IN CRYSTALS

Edited by

N. G. Basov

P. N. Lebedev Physics Institute
Academy of Sciences of the USSR
Moscow, USSR

Translated from Russian by
Albin Tybulewicz

Editor: *Soviet Journal of Quantum Electronics*

CONSULTANTS BUREAU
NEW YORK AND LONDON

Library of Congress Cataloging in Publication Data

Main entry under title:

Luminescence centers in crystals.

(Proceedings (Trudy) of the P. N. Lebedev Physics Institute; v. 79)
Translation of Tsentry lyuminestsentsii v kriśtallakh.
Includes bibliographical references and index.
1. Color centers—Addresses, essays, lectures. 2. Crystal optics—Addresses, essays,
lectures. 3. Luminescence—Addresses, essays, lectures. I. Basov, Nikolaĭ Gennadievich,
1922-   II. Series: Akademiĭa nauk SSSR. Fizicheskiĭ institut. Proceedings; v. 79.
QC176.8.06T7313   548'.9   76-41167

ISBN 978-1-4684-8886-9   ISBN 978-1-4684-8884-5 (eBook)
DOI 10.1007/978-1-4684-8884-5

The original Russian text was published by Nauka Press in Moscow in 1974 for the
Academy of Sciences of the USSR as Volume 79 of the Proceedings of the P. N. Lebedev
Physics Institute. This translation is published under an agreement with the Copyright
Agency of the USSR (VAAP).

PREFACE

This collection of papers describes investigations of luminescence centers in II—VI crystal phosphors, ruby, and molecular crystals. These investigations were carried out using spectroscopy in a wide range of wavelengths, electron paramagnetic resonance, and polarization methods. The relationship between the thermal and optical depths of electron traps is considered specifically.

The articles in this collection should be of interest to all scientists investigating the luminescence of solids.

CONTENTS

Luminescence Polarization Investigation of the Structure of
Doped Molecular Crystals
N. D. Zhevandrov and T. V. Il'inykh

Influence of a Strong Electromagnetic Field on Phase
Transitions in Ferroelectrics
B. P. Kirsanov

INVESTIGATION OF THE SPECTRAL-LUMINESCENCE PROPERTIES OF RUBY AS ACTIVE LASER MEDIUM*

V. B. Neustruev

Investigation was made of the spectra of Cr^{3+} in Al_2O_3 in the 900-8000 Å range. A study was made of the nature of the color centers formed in ruby by high-power optical and γ-ray irradiation. The influence of an anisotropic perturbation by an external electric field on the probability of some transitions in the Cr^{3+} ion was determined. The results obtained made it possible to give a qualitative interpretation of the Cr^{3+} spectrum in the vacuum ultraviolet region, refine the mechanism of quantum losses in the photoexcitation of ruby, and propose a model of the energy band structure. The influence of the color centers on the Cr^{3+} spectrum and on the laser characteristics was considered.

INTRODUCTION

Ruby has been attracting investigations for a long time and its spectral-luminescence properties in the visible range have been investigated in detail and interpreted in the late 1950s. However, the construction of the first ruby laser and the subsequent explosive growth of quantum electronics have raised many theoretical and practical problems which have required further spectroscopic and luminescence investigations of this material under the special conditions of high-power optical excitation. The first to be tackled were the properties of great practical importance and closely related to the current problems in solid-state physics, including the stability and controllability of the stimulated radiation characteristics and optimization of the optical pumping conditions so as to achieve the maximum laser efficiency. These problems made it necessary to review the available results and to fill the gaps relating to the energy spectrum of the activator centers, probabilities of intracenter transitions, and role of recombination processes under strong optical excitation conditions.

The present investigation was concerned with the properties of local Cr^{3+} centers in Al_2O_3, which were investigated by spectroscopic and luminescence methods in the wavelength range 900-8000 Å, and with the processes responsible for changes in the optical properties of ruby as a result of high-power optical excitation. New results were obtained on the intracenter

---

* This paper is based on the thesis submitted for the Degree of Candidate of Physicomathematical Sciences defended at the Lebedev Physics Institute, Moscow on February 28, 1972. The work was carried out under the direction of State Prize Laureate Z. L. Morgenshtern.

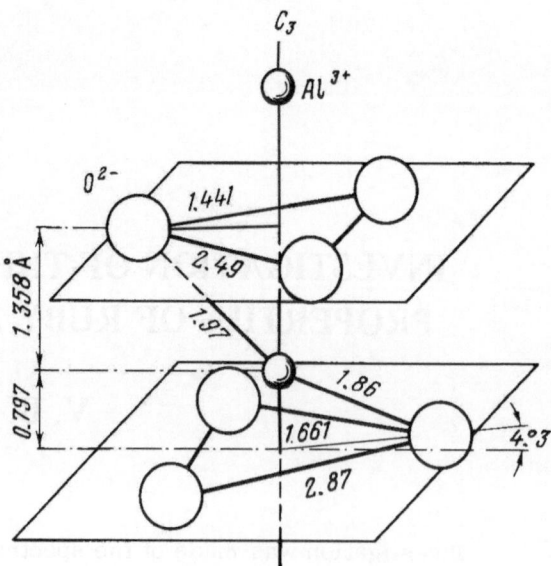

Fig. 1. Position of the Al^{3+} ion in the corundum lattice. The number alongside the
lines are the distances in angstroms.

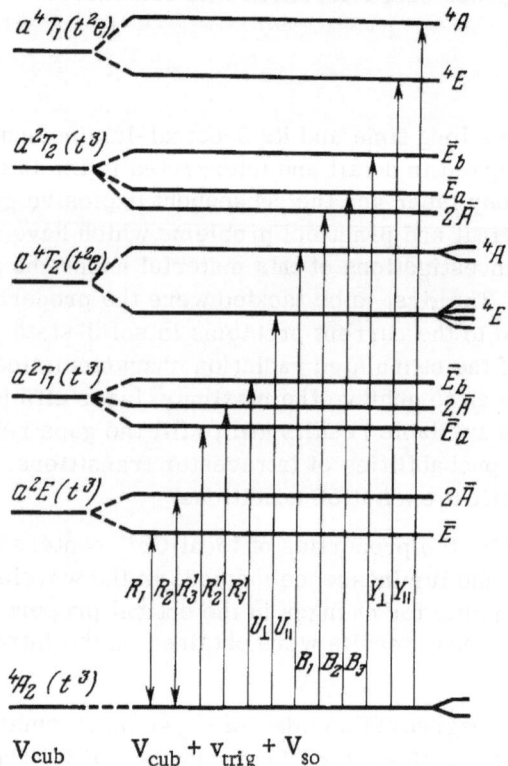

Fig. 2. Energy level scheme of Cr^{3+} in
ruby. The splitting in a field V_{cub} is shown
on the left and the splitting in a combined
field $V_{cub} + V_{trig} + V_{so}$ is shown on the right
[2, 3].

processes, mechanism of the formation of color centers as a result of high-power optical excitation was determined, properties of the color centers and their influence on the laser characteristics were studied, an energy-band model was proposed, influence of the color centers on
the probability of intercombination transitions was discovered, and practical recommendations
for the stabilization of the ruby laser characteristics were deduced.

Ruby ($Al_2O_3 : Cr^{3+}$) is the α modification of corundum containing Cr^{3+} ions which have
replaced isomorphously Al^{3+} ions. An Al^{3+} ion is surrounded by six O^{2-} ions which form a
distorted octahedron (Fig. 1). The cation position has the point symmetry group C_3, but because the symmetry plane is only slightly disturbed, we can regard the cation point symmetry
as C_{3v}. A Cr^{3+} ion is shifted, compared with the Al^{3+} ion, by 0.1 Å along the C_3 axis in the
direction of an octahedral void [1]. The four cations in the unit cell occupy energetically
equivalent positions which are doubly orientation-degenerate.

Pure Al_2O_3 is transparent in a wide spectral range up to ~150 nm, and the color of ruby
is due to the activator absorption resulting from induced electric−dipole transitions between
states of the $3d^3$ electron configuration of Cr^{3+} (d−d transitions), whose oscillator strength
is $f \approx 10^{-4}$. Figure 2 shows the spectrum of Cr^{3+} ($3d^3$ configuration) in ruby calculated allowing for the coexistence of the cubic V_{cub} and trigonal V_{trig} components of the spin−orbit interaction V_{so}. Table 1 gives the results of experimental investigations of the absorption spectra
of ruby.

The luminescence spectrum of ruby consists mainly of narrow R lines (6929 and 6943 Å
at 300°K) associated with the $^2E \rightarrow {}^4A_2$ resonance transition, which is also the active laser
transition. The luminescence of ruby in the 700-760 nm range is due to exchange-coupled Cr^{3+}
pairs but the interpretation of this spectrum is not yet complete.

The intracenter optical excitation of the luminescence (Fig. 3) consists of successive
excitation from the ground state to one of the levels of the quartet ($^4T_2$, a^4T_1, b^4T_1) or doublet
(^2E, $^2T_1$, $^2T_2$, etc.) systems, effective nonradiative relaxation from the excited state to a metastable ^2E level, and radiative decay of the ^2E level characterized by a time constant $\tau \approx 3.6 \cdot 10^{-3}$
sec (300°K, 0.05 at. % Cr^{3+}). The time constant of the nonradiative relaxation to the ^2E state
does not exceed ~10^{-10} sec [8, 9]. The extremely high quantum efficiency of the luminescence
η in the excitation of intercombination transitions [10, 11] and studies of the "excited absorption" [12] indicate that there are no quantum losses in the relaxation of the excitation in the
doublet level system. According to theoretical estimates, the probability of nonradiative decay of the ^2E state is 1.4 sec^{-1} [13].

Calculations of the spectrum of Cr^{3+} in the α-Al_2O_3 lattice, carried out using the crystal
field theory, have made it possible to interpret qualitatively the Cr^{3+} spectrum, but there are
some basic discrepancies indicating that it is not sufficient to allow just for the intraion interactions. For example, the initial splitting of the states $^4A_2$, ^2E, $^2T_1$, and $^2T_2$ cannot be explained
simply by considering V_{trig} and V_{so}. The crystal field theory does not deal with the intensities
of the forbidden transitions and optical anisotropy. The problem of intense intercombination

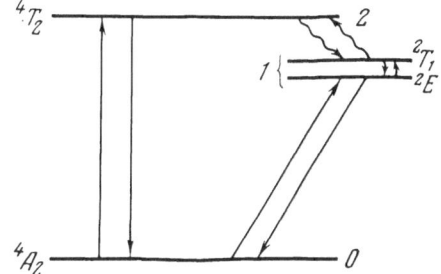

Fig. 3. Luminescence excitation scheme for
ruby.

V. B. NEUSTRUEV

TABLE 1. Positions of Band Maxima Due to Absorption by Isolated Cr^{3+} Ions in Ruby

Band designation	Electron transition	300° K σ ν_{max}, cm^{-1}	300° K σ f_σ	300° K π ν_{max}, cm^{-1}	300° K π f_π	Low temperature σ ν_{max}, cm^{-1}	Low temperature σ f_σ	Low temperature π ν_{max}, cm^{-1}	Low temperature π f_π	Remarks
R_1	$^4A_2(t_2^3) \to E$ } $a^2E(\tfrac{3}{2})$	14 398	$1.5 \cdot 10^{-6}$	14 398	$0.03 f_\sigma$	14 418	$1.5 \cdot 10^{-6}$	14 418	$0.03 f_\sigma$	77° K [4]
R_2	$\to 2A$	14 427	$0.8 \cdot 10^{-6}$	14 427	$0.16 f_\sigma$	14 447		14 447	$0.16 f_\sigma$	
R_1'	$^4A_2(t_2^3) \to E_a$ } $a^2T_1(t_2^3)$	—	—	—	—	15 190		15 190		
R_2'	$\to 2A$	—		—		15 168		15 168		1,8° K [5]
R_3'	E_b	—		—		14 957		14 957		
B_1	$^4A_2(t_2^3) \to 2A$ } $a^2T_2(t_2^3)$	20 973		20 973		20 993		20 993		
B_2	$\to E_a$	21 039		21 039		21 068		21 068		4.2° K [6]
B_3	E_b	21 340		21 340		21 352		21 352		
K_{11}	$E \to 2A$ } $a^2E(\tfrac{3}{2}) \cdots a^2T_2(t_2^3)$	—	—	—	—	6 576	$1 \cdot 10^{-5}$	6 650	0	
K_{12}	$E \to E_a$						0	6 546	$1 \cdot 10^{-5}$	95° K [7]
K_{21}	$2A \to 2A$					6 620	$1 \cdot 10^{-5}$		$1 \cdot 10^{-5}$	
K_{22}	$2A \to E_a$						0		0	
U	$^4A_2(t_2^3) \to {}^4E$ } $^4T_2(t_2^2e)$	18 000	$4.8 \cdot 10^{-4}$	18 450	$1.3 \cdot 10^{-4}$	18 000	$4.8 \cdot 10^{-4}$	18 450	$1.3 \cdot 10^{-4}$	
Y	$^4A_2(t_2^3) \to {}^4E$ } $a^4T_1(t_2^2e)$; ^4A	24 400	$5.88 \cdot 10^{-4}$	25 200	$10.16 \cdot 10^{-4}$	24 400	$5.88 \cdot 10^{-4}$	25 200	$10.16 \cdot 10^{-4}$	5° K [1]
V	$^4A_2(t_2^3) \to b^4T_1(t_2e^2)$	39 000		39 000		39 000		39 000		

Notation: f_σ and f_π are the oscillator strengths in the $E \perp C_3$ and $E \parallel C_3$ configurations, respectively.

transitions is of special importance. The current approach to this problem in the case of ruby is based on a hypothesis put forward by Sugano and Tanabe [2] that the dependence of the probability of intercombination transitions ($^4A_2, \rightarrow \, ^2E, \, ^2T_1$ and $^2T_2$) on the intensity of the transitions between the quartet levels ($^4A_2 \rightarrow \, ^4T_2, \, a^4T_1$) is linear because of the weak spin−orbit interaction. In particular, it is usual to assume that the intensity of the $^4A_2 \rightarrow \, ^2E$ transition depends on the intensity of the U band ($^4A_2 \rightarrow \, ^4T_2$, $\lambda_{max} \simeq 550$ nm). Our results suggest that a review of this viewpoint is needed. There are also some special problems which we shall hope to solve.

CHAPTER I

SPECTRAL DISTRIBUTION OF LUMINESCENCE QUANTUM EFFICIENCY AND ABSORPTION SPECTRA

§ 1. Spectra of Ruby in the 900−2500 Å

Range [14, 15]

The spectra were investigated in order to obtain information on the forbidden-band width and the photoionization limit of the activator centers. The ultraviolet spectra were investigated using photometric apparatus (with a sensitivity of $\sim 10^3$ photons/sec) based on a vacuum monochromator with normal incidence and grating dispersions 8 and 30 Å/mm. The spectral distribution of the quantum efficiency η (λ) was determined by the integrating sphere method employing sodium salicylate to convert the ultraviolet radiation to the visible wavelengths. Photomultipliers calibrated to within 1% were used as detectors. The main assumption made in the calculation of η (λ) was the constancy of the quantum efficiency of sodium salicylate in the 900−3000 Å range [16].

Our investigation of the absorption spectra (obtained using unpolarized light at 300°K) of ruby revealed a strong wide band at $\lambda \lesssim 2100$ Å overlapping the fundamental absorption region of corundum (Fig. 4). This short-wavelength activator absorption band at $\lambda <$ 2100 Å was called the K band. The absorption coefficient $\varkappa$ of this band could be measured only for corundum samples with traces of chromium (C < 0.005 at. %, curve 2 in Fig. 4).* The K band showed no structure when the temperature was lowered to 80°K. At the maximum of this band the absorption cross section was $\sigma_{max} \geqslant 4 \cdot 10^{-17}$ cm^2 (h$\nu_{max} \approx 6.9$ eV), and the oscillator strength was $f \geqslant 0.3$. Somewhat later this band was investigated by Loh [18], and he found that the K band was unpolarized with $\sigma_{max} \approx 7 \cdot 10^{-17}$ cm^2 and $f \approx 0.3$.

Excitation with light corresponding to the K band generated strong luminescence with the same spectrum as that obtained as a result of excitation in the U and Y bands. However, the quantum efficiency of the luminescence resulting from the excitation in the K-band region was higher than in the case of excitation in the U and Y bands, and the value of η (K) for samples with C(Cr^{3+}) = 0.01−0.13 at. % was 0.8−0.95 (Fig. 5). The spectral distribution of η (K) had a structure with at least six bands; the short-wavelength fall of η (K) at $\lambda <$ 1650 Å coincided with the rise of the fundamental absorption in corundum. However, this did not indicate that

---

* The concentration of Cr^{3+} in ruby was found by an optical method using the empirical formula [17]

$$C \, (\text{wt. } \% \, Cr^{3+}) = 0.29 \frac{D_{555} - D_{700}}{d} \, ,$$

where D_{555} and D_{700} are the densities of a sample d (mm) thick measured in the $E \perp C_3$ case at λ = 555 and 700 nm.

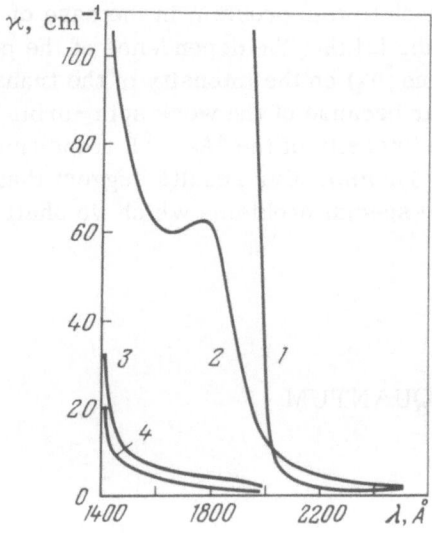

Fig. 4. Absorption spectra of corundum and ruby: 1) ruby, 0.05 at.% Cr^{3+}; 2) corundum with traces of chromium (less than 0.005 at.% Cr^{3+}); 3, 4) corundum.

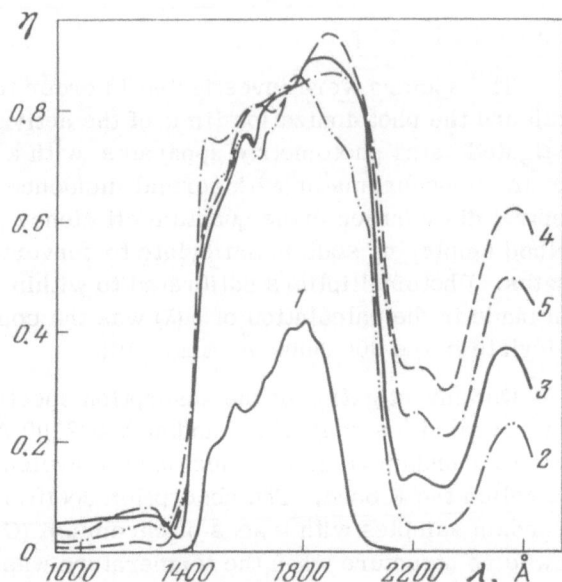

Fig. 5. Influence of the impurity concentration on $\eta(\lambda)$: 1) corundum with traces of chromium (less than 0.005 at.% Cr^{3+}); 2) ruby with 0.04 at.%; 3) ruby with 0.05 at.% (standard sample); 4) ruby with 0.09 at.%; 5) ruby with 0.12 at.%.

the transfer of the excitation from the lattice to the luminescence centers was inefficient because the absorption coefficient in the $\lambda < 1400$ Å range was $\varkappa \sim 10^5$ cm^{-1} [19], and the influence of the surface became considerable.

The concentration dependences of the spectra in the V-band region [$\lambda_{max} = 2570$ Å, $^4A_2(t_2^3) \rightarrow b^4T_1(t_2e^2)$] were of particular interest. The transition responsible for this band resulted in a considerable change in the electron orbits (two-electron transition), and theoretically the intensity of the corresponding band should be close to zero. The spectra $\varkappa(\lambda)$ and $\eta(\lambda)$ obtained in the V-band region differed from one ruby crystal to another, and this gave rise to different views on the nature of the V band. It should be possible to determine the nature of this band by investigating the $\varkappa(V)$ and $\eta(V)$ spectra for different concentrations of Cr^{3+}. It is clear from Fig. 5 that $\eta(V)$ varied strongly with the concentration $C(Cr^{3+})$. However, if it was assumed that this was due to inactive absorption of the host lattice, competing with the activator absorption $\varkappa_0(V) = k_0C$, and the inactive absorption was subtracted, the true characteristics of the V band were obtained, and these were in agreement with the experimental results (Table 2) and with the requirement that $\eta_0(V)$ should be independent of the concentration.

TABLE 2. Experimental Values of
Luminescence Efficiency and Absorption
at V-Band Maximum (λ_{max} = 2570 Å)

Crystal No.	c, at. %	$\varkappa$, cm⁻¹	η	Curve in Fig. 5
1	0.01	0.15	0.4	—
2	0.04	0.74	0.23	2
3	0.05	0.62	0.41	3
4	0.09	0.78	0.62	4
5	0.12	1.26	0.50	5

The values of $\eta_0(V)$ and k_0 at the V-band maximum were 0.65 and 8 cm⁻¹/at. %. This confirmed the relationship between the V band and the intracenter transition in Cr^{3+}, and it also demonstrated that the inactive absorption played a considerable role in this part of the spectrum. The results of an investigation of the spectra in the K-band region [η (K) $\approx$ 1, quantum efficiency independent of the impurity concentrations, and absorption characteristics] also supported the intracenter nature of the transitions in this part of the spectrum. The spectrum of ruby in the $h\nu > 4 \cdot 10^4$-cm⁻¹ range has frequently been called the charge-transfer region. However, this term should not be understood literally. Ionization of impurities or interband transitions would have manifested themselves in the dependences of $\eta(\nu)$ on the impurity concentration and on the exciting radiation.*

The K-band absorption (48,000-72,000 cm⁻¹) cannot be attributed to transitions in the quartet system of the Cr^{3+} levels, whereas in the energy range 45,000-74,000 cm⁻¹ there are eight excited states of the doublet system ranging from c^2T_2 to d^2E [20] (Fig. 6), and we may assume that the transitions responsible for the K-band absorption terminate at these levels. Intercombination transitions are subject to a double forbiddenness in respect of the parity and spin but nevertheless the intensity of the K-band absorption suggests that electric-dipole transitions may be allowed in practice. This conflict is removed if we bear in mind that the part of the spectrum under consideration is located near the fundamental absorption edge and near the Cr^{3+} states of opposite parity ($3d^24p$ configurations). Mixing of the higher excited states of the doublet system of Cr^{3+} with the conduction-band states and with the $3d^24p$ configuration of Cr^{3+} should violate the selection rules, and this should increase the intensity of the intercombination transitions.

The absence of structure in the dependence $\varkappa(\lambda)$ in the K-band region can also be explained quite easily. The upper excited states of the doublet system are genetically related to the terms of the quartet system with the t_2^2e and t_2e^2 configurations, and, therefore, the transitions from the ground state (t_2^3 configuration) to the upper doublet states result in changes in the electron shells and should be manifested as absorption bands $10D_q \approx 2000$ cm⁻¹ wide (D_q is the cubic field parameter) [21]. Moreover, a calculation of the Cr^{3+} spectrum given in [20] is made in the cubic field approximation so that if we allow for the high degree of degeneracy (4.6) or the $\{c^2T_2...d^2E\}$ states, we can expect splitting of each state in the trigonal field by ~ 500 cm⁻¹. Thus, the K absorption band attributed by us to the intercombination $^4A_2 \rightarrow \{c^2T_2...d^2E\}$ transitions is a superposition of a large number (at least six) of wide bands † with

---

*No special investigation of the dependences of $\eta(\lambda)$ on the excitation intensity was made, but within the limits of the K band the source intensity varied by a factor of several tens and yet there was no correlation between the excitation intensity and $\eta(\lambda)$.

†It is possible that the $\lambda \approx 2250$ Å band in the $\eta(\lambda)$ spectrum (Fig. 5) corresponds to the $^4A_2 \rightarrow c^2T_2$ transition, whereas the $^4A_2 \rightarrow e^2T_1$, d^2E transitions (calculated position 1350 Å) overlap completely the fundamental absorption region.

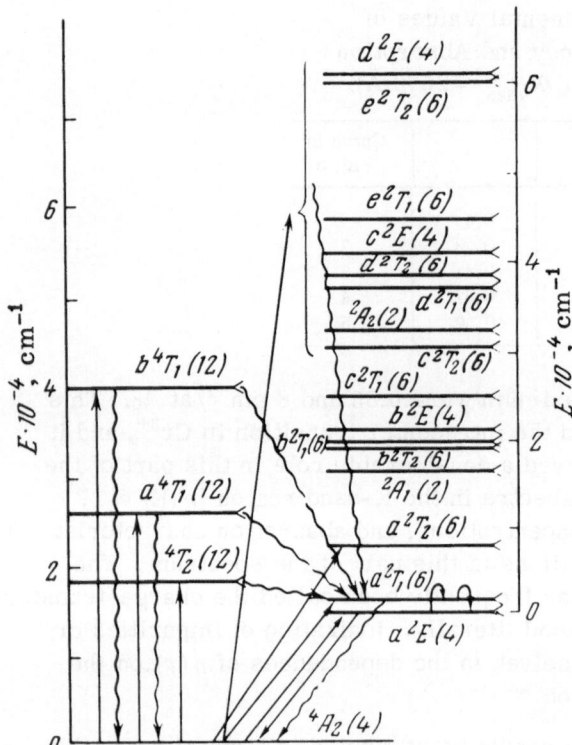

Fig. 6. Energy level scheme of Cr^{3+} in ruby.

$f \sim 0.05$. An investigation of the excited absorption in the region corresponding to the $^2E(t_2^3) \rightarrow \{c^2T_2...d^2E\}$ transitions has established the presence of a wide absorption band with a weak structure ($T \geq 40°K$) [22].

The structure of the K band is manifested most clearly in the $\eta(\lambda)$ spectrum (Fig. 5), where even a superficial inspection reveals six components of curves 1 and 2 and four components (three long-wavelength components merge into one band) of curves 3-5. The K-band structure can be attributed to the different values of the quantum losses associated with the origin of the excited states because the relationship of each of the $\{c^2T_2...d^2E\}$ terms with the quartet levels is different.

§2. Temperature Dependence of Luminescence
Quantum Efficiency under Resonance Excitation
Conditions [23, 24]

Any luminescence excitation process in ruby terminates with the radiative transition $^2E \rightarrow ^4A_2$ so that the losses experienced during this transition govern the upper limit of the luminescence efficiency. It has been shown in several papers that the efficiency of the excitation of the resonance transition $\eta(R)$ is close to unity at $T = 300°K$. However, in the foreign (non-Soviet) literature there is a widely held incorrect view that the value of $\eta(R)$ is unity in the range $T < 300°K$ and that this has been established experimentally.* The value $\eta(R) = 0.94 \pm 0.05$ at $300°K$ is reported in [10]. However, the mechanisms of the luminescence quantum cannot be determined without an accurate knowledge of the value of $\eta(R)$ and its tem-

---

*Usually a reference is made to Maiman's paper [25], who deduces that $\eta(R)$ is close to unity on the basis of an approximate comparison of the probabilities of transitions in a three-level system.

perature dependence. Moreover, our investigation of the luminescence intensity under the influence of an external electric field yielded results indicating that $\eta(R)$ at 77°K differed considerably from unity so that it was necessary to investigate the temperature dependence of $\eta(R)$. We developed a method for precision measurements utilizing an integrating sphere, and this enabled us to find $\eta(R)$ to within 1% in the temperature range 80-300°K.

In earlier studies the experimental error in the determination of $\eta(R)$ was mainly due to the inaccuracy of the separation of the scattered exciting light from the luminescence. In our method this separation was not made, and the main source of error was thus removed. The high efficiency $\eta(R)$ enabled us to make an accurate direct allowance for the luminescence quantum losses, which were deduced from the difference between the sum of all the light fluxes (luminescence, light transmitted by a crystal, and reflected light) and the exciting light flux at the resonance transition wavelength.

A detailed description of the method and an analysis of the experimental errors are given in [23]. Here we shall simply mention that $\eta(R)$ is described by the formula

$$\eta(R) = \frac{S_R}{S_\lambda} \left(\frac{\sigma_i \{ r[1 + (1-r)^2] + p(1-r)^2\} - (1-r)^2 \sigma_t p + r[1 + (1-r)^2 \sigma_t^2]}{(1-r)[1 - r(1-r)\sigma_t^2 - (1-r)\sigma_t]} \right), \qquad (1)$$

where σ_i and σ_t are the measured quantities representing the ratio of the sum of the fluxes to the incident flux (σ_i) and to the transmitted flux (σ_t); r is the specular reflection coefficient of ruby; p is the diffuse reflection coefficient of the MgO-coated screen which scatters the light transmitted by a sample; S_R/S_λ is the ratio of the sensitivities of a photomultiplier to the exciting light and to the luminescence. The parameter p is assumed to be 0.99, and possible deviations of p from this calculated value give rise to a methodological error in $\eta(R)$ which is estimated to be 1%. In practice, calculations are made using the formula

$$\eta(R) = \frac{S_R}{S_\lambda} \left(\frac{\sigma_1 - \sigma_t}{1 - \sigma_t} + \Delta\eta \right), \qquad (2)$$

where $\Delta\eta$ is a small correction which depends on σ_i and σ_t.

The temperature dependences of $\eta(R)$ are plotted in Fig. 7. The spectral slit width in the measurements of these dependences was 1 Å but the low-temperature spectral distortions did not affect the precision of the measurements if $\sigma_t \lesssim 0.5$. It is clear from Fig. 7 that $\eta(R) = 1.00 \pm 0.01$ only for some crystals in the temperature range 140-200°K. In the case of colored crystals (curves 4 and 6) and of a crystal with a high activator concentration (curve 2) the value of $\eta(R)$ was always 6-8% less than unity. The high precision of our measurements established that the luminescence efficiency for the excitation of the R_2 line, $\eta(R_2)$, was either equal to $\eta(R_1)$ or it was 0.005-0.01 lower in the investigated temperature range.

The high-temperature fall of the efficiency at $T > 200°K$, exhibited by all samples, could be approximated satisfactorily by the formula

$$\eta(R) = \frac{W_{10}^r}{W_{10}^r + W_{40} \exp(-\Delta E_{14}/kT)} \qquad (3)$$

with the quenching parameters $\Delta E_{14} = 0.08 \pm 0.03$ eV and $W_{40}/W_{10}^r = 0.1$ (Fig. 8). The value of ΔE_{14} represents the position of the quenching level relative to the ^2E state and W_{40} and W_{10}^r represent the probabilities of nonradiative and radiative decay of the ^2E state. The results obtained indicate that the initial stage of the thermal quenching involves not the $^4T_2$ state which is located 0.30 eV above ^2E, but the nearest doublet level $^2T_1$ located 0.07 eV above ^2E. An estimate of the probability of nonradiative decay of the $^2T_1$ state gives $W_{40} \approx 30$ sec^{-1}, which

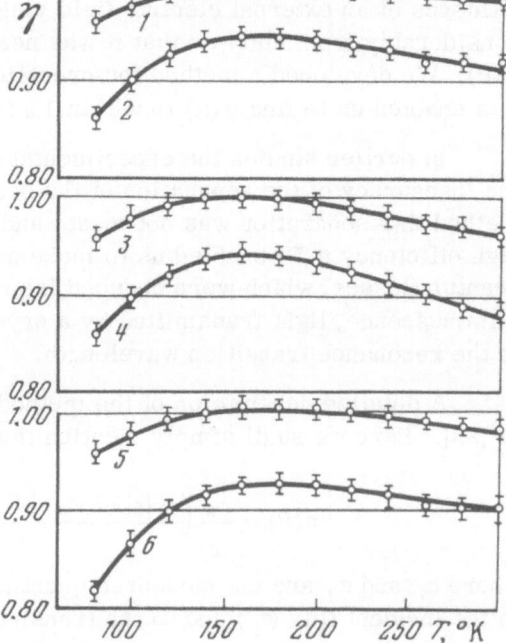

Fig. 7. Temperature dependences of η (R): 1) high-quality laser ruby crystal; 2) ruby with high activator concentration (0.13%); 3, 4) laser rubies cut from the same rod (3 – uncolored crystal, 4 – optically colored crystal); 5, 6) laser rubies cut from the same rod (5 – uncolored crystal, 6 – crystal colored by γ-ray irradiation).

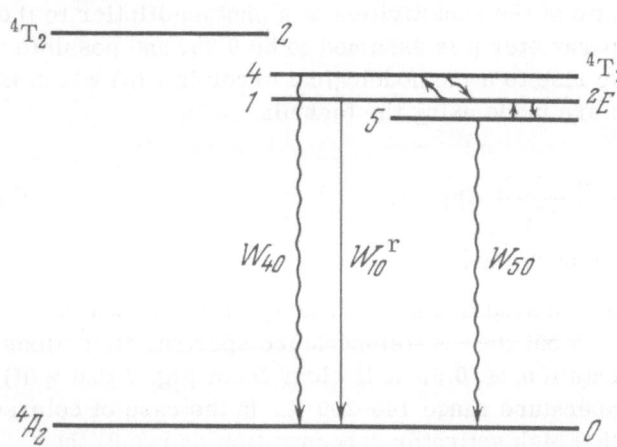

Fig. 8. Thermal quenching of η (R).

is more than an order of magnitude higher than the probability of nonradiative decay of the ^2E state ($W_{10} \approx 1.4$ sec^{-1}). The ratio W_{10}/W_{40} is equal to the ratio of the widths of the absorption lines resulting from the $^4A_2(t_2^3) \rightarrow {}^2E(t_2^3)$ and $^4A_2(t_2^3) \rightarrow {}^2T_1(t_2^3)$ transitions. The stronger the interaction between the doublet state with the quarter or ligand states, the greater is the width of the absorption band of the corresponding intercombination transition and the greater is the probability of this transition. Thus, our value of $W_{40} \approx 30$ sec^{-1} for the $^2T_1 \rightarrow {}^4A_2$ nonradiative transition is in agreement with the lifting of the forbiddenness of the intercombination transitions in ruby.

The low-temperature fall η (R) is quite unexpected. Phenomenologically, this fall can be attributed to the presence of a weak quenching level (level 5 in Fig. 8), located below ^2E at a depth $\Delta E_{15} \approx 0.02$ eV (160 cm^{-1}). The position of this level does not agree with any Cr^{3+} state. It is possible that the value of ΔE_{15} is the effective depth of weak-quenching levels associated with the exchange interaction of the Cr^{3+} ions because the low-temperature quenching is stronger in ruby crystals with higher Cr^{3+} concentrations.

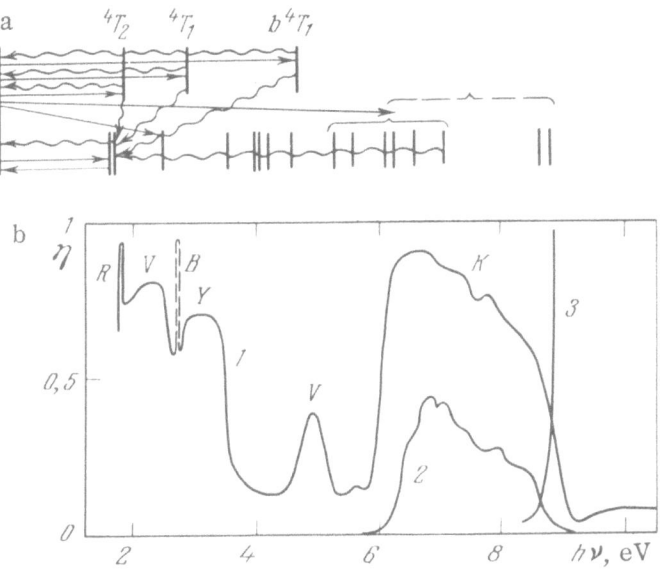

Fig. 9. Electron transitions (a) and spectral distribution of the luminescence quantum efficiency of ruby (b): 1) ruby with 0.05 at.% Cr^{3+}; 2) corundum with traces of chromium (0.005 at.%); 3) fundamental absorption edge of corundum; the R band is located at 2 eV.

We shall now summarize the results of the experimental investigations of $\eta(\lambda)$, as presented in Fig. 9 and Table 3. A comparison of the values of η for the excitation in the quartet and doublet level systems shows that the losses are mainly due to nonradiative transitions between the quartet states. In the doublet system the excitation relaxes to the ^2E level practically without any quantum losses. The slight fall of $\eta(\lambda)$ within the K band is partly due to the inactive absorption of the host lattice, which increases in the direction of shorter valence, and

TABLE 3. Distribution of Luminescence Quantum Efficiency
of Ruby at 300°K

Excitation region	Electron transition	Quantum efficiency	Remarks
U band	$^4A_2(t_2^3) \rightarrow {}^4T_2(t_2^2e)$	0.80	Excitation in quartet level system
Y band	$^4A_2(t_2^3) \rightarrow a^4T_1(t_2^2e)$	0.70	
V band	$^4A_2(t_2^3) \rightarrow b^4T_1(t_2e^2)$	0.65	
R lines	$^4A_2(t_2^3) \rightarrow a^2E(t_2^3)$	0.96	Excitation in doublet level system
B lines	$^4A_2(t_2^3) \rightarrow a^2T_2(t_2^3)$	1 [11]	
K band	$^4A_2(t_2^3) \rightarrow \begin{cases} c^2T_2(t_2^3) \\ {}^2A_2(t_2^2e) \\ d^2T_1(t_2^3) \\ d^2T_2(t_2^3) \\ c^2E(t_2^3) \\ e^2T_1(t_2^3) \\ e^2T_2(t_2^3),\ d^2E(t_2^3) \end{cases}$	0.95 Maximum	

partly due to the origin of the states $\{c^2 T_2 ... d^2 E\}$. The dominant relaxation channel in the case of excitation in the K-band region remains the nonradiative transition to the ^2E state. Quantum losses in the excitation of the doublet system at $T \lesssim 300°K$ are mainly due to the quenching via the $^2T_1$ level whereas at $T \lesssim 140°K$ they are due to the quenching via exchange-coupled states of the Cr^{3+} pairs. At $T > 400°K$, the predominant nonradiative decay channel of the ^2E state is represented by the transitions involving $^4T_2$.

CHAPTER II

PHOSPHORESCENCE OF RUBY. ENERGY BAND SCHEME

Long-lasting afterglow (phosphorescence) emitted by optically excited ruby and corundum crystals has been attributed to crystal defects. We discovered phosphorescence which appeared after high-power optical excitation which was exhibited only by ruby crystals. We established first the identity of the phosphorescence and photoluminescence spectra which indicated that the luminescence centers were chromium ions and not accidental impurities or lattice defects. This was investigated in detail because the excitation of phosphorescence during stimulated emission could make a considerable contribution to the energy balance of a laser. Moreover, the determination of the phosphorescence excitation mechanism was of intrinsic interest.

§ 1. Mechanism of the Excitation of Ruby

Phosphorescence

Phosphorescence of ruby crystals was investigated by pulse photoexcitation at room temperature [26]. The phosphorescence intensity was measured using an interference filter (696 ± 5 nm) 0.2 sec after the end of excitation. A sample was illuminated in a laser enclosure by a single flash produced by a pulse-discharge lamp. The phosphorescence decay in the period $2 \text{ sec} < t < 10^3 \text{ sec}$ could be approximated satisfactorily by a first-order hyperbola. The initial phosphorescence brightness (extrapolated values) increased superquadratically with the pump power, i.e., the exponent in the dependence was ~ 2.5 ($\lambda_{exc} = 446$ nm). This behavior indicated, on the one hand, that the recombination process was bimolecular and, on the other hand, that the excitation process was not simple [27]. The recombination luminescence spectrum observed for all the crystals clearly indicated a relationship with the photoionization of the Cr^{3+} ions as a result of optical excitation followed by the recombination of free electrons at the Cr^{4+} ions formed in this way.

The phosphorescence excitation mechanism was determined by investigating the excitation spectrum in the 210-700 nm range. At wavelengths of 400-700 nm a sample was excited through interference filters with a transmission half-width $\Delta\lambda = 10$ nm, whereas in the range 210-400 nm the exciting light was passed through a high-luminosity monochromator with a spectral slit width of 8 nm. The relative efficiency of the phosphorescence excitation q was defined as the ratio of the quantities proportional to the number of the phosphorescence photons to the number of the absorbed exciting-light photons:

$$q(\lambda) = \frac{\int_0^\infty I(t)\, dt}{E(\lambda)\, \Delta\lambda\, (1 - P)}, \tag{4}$$

where $E(\lambda)$ is the excitation intensity and P is the transmission of a crystal at $\lambda = \lambda_{exc}$. The dependence $q(\lambda)$, plotted in Fig. 10, gives a qualitative idea on the excitation efficiency because

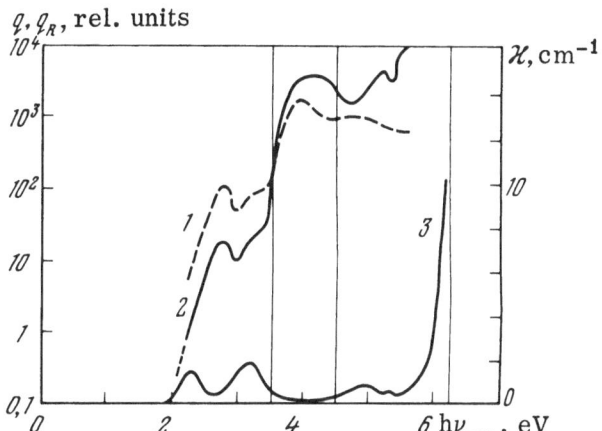

Fig. 10. Efficiency of the excitation of ruby phosphorescence: 1) excitation spectrum $q(\nu)$; 2) excitation efficiency $q_R(\nu)$; 3) absorption spectrum of ruby.

firstly, this process is strongly nonlinear and, secondly, integration is carried out over a short time interval. An estimate of the excitation efficiency obtained for the absorption of 10^{16} photons ($\lambda_{exc} = 446$ nm) gives $q \sim 10^{-7}$. Since a further increase in the excitation intensity causes the phosphorescence to rise subquadratically, it follows that the recombination processes do not make a significant contribution to the energy balance of a laser.

The phosphorescence spectrum is correlated with the activator absorption spectrum, and this confirms the hypothesis of the photoionization of the Cr^{3+} ions. The long-wavelength phosphorescence excitation edge is located at 603 nm (≈ 2 eV). This photon energy is insufficient for the one-photon ionization of the Cr^{3+} ion because this requires at least 6.3 eV. Moreover, this energy is insufficient for a one-photon optical transition from the ^2E state to the conduction band ($6.3 - 1.78$ eV $= 4.5$ eV), but nevertheless the cascade photoionization via the ^2E level is most likely. We may assume that the multiphoton ionization process does not make a significant contribution when the rate of excitation is $\sim 10^{16}$ photons/sec.

The nature of the photoionization was determined by exciting phosphorescence with two pulses separated by a variable time interval [26]. One light pulse (lasting 2.2 msec) excited ruby at $\lambda_{exc} = 696 \pm 5$ nm (corresponding to the R lines) or at 549 ± 5 nm; the second (delayed) pulse was of 0.8 msec duration, and it excited a crystal through a monochromator in the spectral range 210-580 nm. When the delay time was 2.4 msec, so that the pulses were completely separate, and the population of the ^2E level after the excitation of the first pulse was still fairly high, we observed nonadditive excitation of the phosphorescence, i.e., the integrated phosphorescence intensity under the action of two consecutive pulses was greater than the sum of the phosphorescence intensities obtained for each pulse separately. This proved the cascade mechanism of the photoionization of Cr^{3+} in which the ^2E level was the intermediate stage.

The mechanism of the phosphorescence excitation in ruby can be described as follows. High-power optical excitation produces a strong occupancy of the metastable ^2E state from which excitation passes to the upper states of the doublet system and then relaxes ending with the $^2E \rightarrow ^4A_2$ radiative transition. Only a small proportion of the excited Cr^{3+} ions is ionized either by the direct transitions from the ^2E level to the "conduction band" or as a result of tunneling from the excited states in the doublet system. The correlation between the excited absorption spectrum and the $q(\lambda)$ spectrum, as well as the structure of $q(\lambda)$, suggest that the tunneling of electrons is more likely. In the case of indirect optical transitions from the ^2E level to the band we may expect a monotonic dependence $q(\lambda)$ with rising $h\nu_{exc}$: $q(\nu) \sim (h\nu - E_T)^n$, where n = 0.5-2 [28]. Further studies of this point would be desirable but since it did not affect subsequent conclusions, it was assumed specifically that the ionization of Cr^{3+} was due to the tunneling from the excited states in the doublet system.

§ 2. Energy-Band Model of Ruby

The excitation of phosphorescence by $h\nu = 2$ eV quanta in a crystal with a wide forbidden band demands an assumption that the energy bands of corundum are bent in the **k** space. The theory of the energy bands of ionic crystals is still being developed, and the representations should be applied to ionic crystals with caution. However, experience shows that the band model of ionic crystals provides a useful description of the recombination processes and of the phenomena associated with the ionization of impurity centers and of the host lattice.

The band model of ruby has been developed using the results obtained in studies of ultraviolet spectra of ruby and corundum and of the phosphorescence excitation spectra. The main assumptions made are associated with the optical forbidden-band width E_0 and with the position of the ground state of Cr^{3+} in the forbidden band. The absorption coefficient of corundum rises strongly in the region $\lambda \leq 1400$ Å. This rise has been attributed to the onset of direct transitions to the conduction band, and, consequently, E_0 has been assumed to be 8.8 eV. A strong rise of $\varkappa(\lambda)$ or ruby is shifted by 2.7 eV compared with corundum. If it is assumed that the long-wavelength edge of the K band corresponds to the onset of the direct transitions between $^4A_2$ level and the "conduction band," the energy 2.7 eV should correspond to the position of the $^4A_2$ level above the top of the valence band. If the criterion of the onset of impurity ionization is the fall of η in the K-band region at $h\nu \geq 6.5$ eV ($\lambda \leq 1900$ Å), the $^4A_2$ level is found to be 2.3 eV above the top of the valence band. In the band model of ruby we shall use the average value of 2.5 eV. Having thus determined the position of the ground state of Cr^{3+} in the forbidden band, we can deduce the position of the conduction-band minimum relative to the $^4A_2$ level. When allowance is made for the long-wavelength phosphorescence excitation edge, it is found that the minimum of the conduction band should be $1.78 + 2.05$ eV ≈ 3.8 eV above $^4A_2$ level, and the thermal width of the forbidden band should be $E_T = 3.8 + 2.5$ eV $= 6.3$ eV. Figure 11 shows the energy-band scheme of ruby and the Cr^{3+} levels. The doublet system levels are displaced arbitrarily along the k axis in order to avoid confusion.

Investigations of the electrical conductivity of corundum [29, 30] have yielded forbidden bands ranging from 8.5 to 11 eV. These values have been calculated by doubling the slope tangent of the dependences $\sigma (1/T)$, i.e., by applying the method for the determination of E_T developed for typical semiconducting materials. If this method is applicable to E_T of corundum, we find that — allowing for the energy shift of the conduction band by 2.5 eV — the optical width of the forbidden band should be within the range 11–13.5 eV so that our value $E_0 = 8.8$ eV is strongly underestimated. A correction to E_0 in the energy-band scheme should therefore take the form of an increase in the energy gap between the $^4A_2$ level and the top of the valence band.

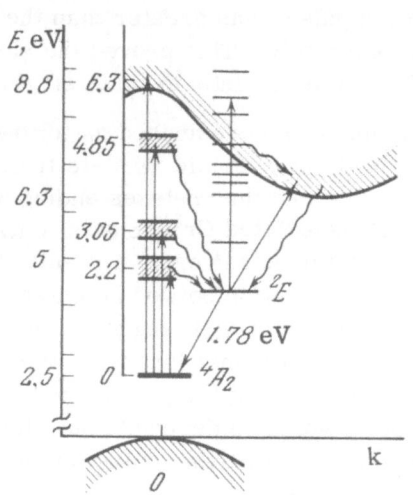

Fig. 11. Energy-band scheme of corundum showing positions of the Cr^{3+} levels (doublet levels of Cr^{3+} are displaced by an arbitrary amount).

It is interesting to note also that the absorption coefficient of corundum at the fundamental edge ($\lambda > 1400$ Å) can be approximated satisfactorily by the function $\varkappa(\lambda) \propto (h\nu - E_T)^2$, which corresponds — in the case of a semiconducting material — to the excitation of indirect transitions. Extrapolation of $\varkappa(\nu)$ to the origin gives $E_T = 5.6 \pm 0.4$ eV.

We shall now analyze the spectral dependence of the phosphorescence excitation efficiency (Fig. 10) bearing in mind the energy-band scheme of ruby. High-power optical excitation with photons of 2 eV $\leq h\nu_{exc} \lesssim 3.5$ eV energy may cause ionization only by a two-stage mechanism in which the second stage represents the transition $^2E \rightarrow \{^2A_1...c^2T_1\} \rightsquigarrow$ "conduction band." In this excitation region the efficiency of electron transfer from the Cr^{3+} ions to the band by excitation pulses shorter than τ is given by

$$q_R(\lambda) \approx \frac{q(\lambda)}{E(\lambda) \, \Delta\lambda\eta(\lambda)} . \tag{5}$$

In the photon energy range 3.5 eV $\lesssim h\nu_{exc} \lesssim 4.5$ eV the ionization process involves simultaneously the multistage mechanism $^2E \rightarrow \{c^2T_1...d^2T_1\} \rightsquigarrow$ "conduction band" and the one-photon mechanism $^4A_2 \rightarrow \{^2A_1...c^2T_1\} \rightsquigarrow$ "conduction band." Indirect one-photon transitions from the ground state to the conduction band are supported by the strong rise of the phosphorescence excitation efficiency (Fig. 10).

In the photon energy range $h\nu_{exc} > 4.5$ eV we may have direct $^2E \rightarrow$ "conduction-band" transitions, but at 4.5 eV ruby is transparent, and the rise of $q_R(\lambda)$ begins only in the V-band region.

Finally, in the range $h\nu_{exc} > 6.3$ eV ($\lambda < 1900$ Å) we may expect one-photon ionization of the Cr^{3+} ions by direct $^4A_2 \rightarrow$ "conduction-band" transitions.

In the range 2.0 eV $< h\nu_{exc} < 5.8$ eV the excitation efficiency q_R rises approximately by a factor of 10^5.

Phosphorescence of ruby was observed after static excitation with $\lambda_{exc} = 161$ nm (strong line in the hydrogen spectrum with $\sim 10^{12}$-10^{13} photons/sec). Phosphorescence was not observed as a result of long-wavelength excitation, but in the range $\lambda > 161$ nm the excitation intensity fell by two orders of magnitude and the luminescence quantum efficiency rose (quantum losses decreased) so that the phosphorescence intensity fell below the sensitivity threshold of the photometric system. Excitation in the wavelength range 900-1400 Å produced phosphorescence even when the intensity of the exciting light was low.

Thus, having established the multistage nature of the phosphorescence excitation by high-power optical radiation we can consider the following consequences of this mechanism:

1. Electrons liberated as a result of the ionization of Cr^{3+} which have not escaped in time from the field of action of a positively charged Cr^{4+} center are recaptured by their "own" center. In this case the recombination process is practically indistinguishable from a purely intracenter process.

2. Liberated electrons escape from the range of action of a Cr^{4+} ion and are localized at shallow electron traps. Thermal liberation of electrons from the shallow traps and subsequent recombination at the Cr^{4+} ions resulting in the emission in the R-line region is responsible for the observed phosphorescence.

3. Some of the liberated electrons are captured by deep traps forming stable color centers. In this case the Cr^{4+} ions should form hole color centers in ruby.

The first two processes are relatively short-lived, and they cause no irreversible change in the optical properties of ruby. The third process should result in the accumulation of color centers, i.e., it should alter the optical properties of a ruby crystal when it is used for a long time as an active laser element.

CHAPTER III

COLOR CENTERS IN RUBY

Changes in the optical properties of a laser element due to the formation of color centers in the course of prolonged operation may degrade the laser characteristics. Studies of the nature of color centers, their optical properties, and interaction with the Cr^{3+} activator ions are the first task in tackling the problem of the stability of ruby laser characteristics.

Much work has been done on color centers in ruby bombarded with x rays, electrons, and γ rays. For example, it follows from the results obtained in [31] that additional absorption appears in colored ruby, and the concentration of Cr^{3+} decreases. This additional absorption disappears when a crystal is heated to $\sim 400°C$. Color centers have been attributed to all possible crystal defects, and the activator impurity has usually been assigned a secondary role. However, the formation of the Cr^{2+} and Cr^{4+} centers in crystals subjected to hard irradiation has been established quite reliably [32, 33].

Studies of the optical coloration of ruby, first observed by Schultz [34], have been few, and they have not established the nature of color centers [35, 36]. It has been assumed in these investigations that chromium plays a secondary role in the formation of color centers. However, the hypothesis of the predominance of lattice defects (those not associated with the Cr^{3+} activator ions) in the formation of color centers seems unconvincing because the concentration of structure defects in ruby crystals of even moderate quality is $\sim 10^{15}$ cm^{-3} and the concentration of accidental impurities is $\sim 10^{16}$-10^{17} cm^{-3}, whereas changes in the concentration of Cr^{3+} after γ-ray coloration may reach $\sim 10^{19}$ cm^{-3} [31]. It follows from this relationship between the concentrations of color centers and defects that the coloration of ruby is mainly due to the changes in the charge of the Cr^{3+} ions. Moreover, it is known that pure corundum is not colored by intense γ irradiation [37]. Hence, it follows that γ irradiation does not produce structure defects which could give rise to color centers.

In establishing the nature of color centers it is very important to compare the optical properties of the centers obtained by different methods. In the case of coloration by hard radiation there is a multiplicity of processes which confuses the situation, but in the case of coloration by high-power optical radiation the problem simplifies considerably. Coloration of ruby by high-power light pulses should be accompanied by the formation of hole Cr^{4+} centers. Localization of electrons at deep traps should produce electron color centers. Since the concentration of color centers is (like changes in the concentration of Cr^{3+}) two orders of magnitude greater than the concentration of other defects, the predominant electron color center may be a Cr^{2+} ion, i.e., a Cr^{3+} activator which has captured an electron. Other defects may also give rise to color centers but this should not be of primary importance because their concentrations are low. The formation of hole O^- centers as a result of optical coloration [31] is unlikely to occur because phosphorescence is excited by an intracenter process in Cr^{3+} ions, and transitions of the $(Cr^{3+}, O^{2-}) + h\nu \rightarrow (Cr^{2+}, O^-)$; and $(Cr^{4+}, O^-) + h\nu \rightarrow (Cr^{3+}, O^-)$ type are not observed in ruby. However, in a comparative investigation of optically and γ-ray colored ruby crystals we made an attempt to discover paramagnetic O^- centers because these were the only possible hole centers (apart from Cr^{4+}) whose formation could explain the presence of color centers in concentrations of $\sim 10^{19}$ cm^{-3}. The ESR spectra of the optically and γ-ray-colored crystals failed to confirm the formation of such centers.

§ 1. Optical Properties of Color Centers

We studied the thermoluminescence, additional absorption ($\Delta\varkappa$), and quantum efficiency spectra; we also studied selective optical bleaching of optically and γ-ray-colored ruby crystals of the type used in lasers [38-40]. In these investigations each rod was cut into three samples; one of the samples was not colored and was used as a control.

Optical coloration was performed by repeated flashing of a pulse-discharge lamp in a laser enclosure. This produced a peripheral orange coloring, and the volume-average concentration of the Cr^{3+} ions decreased by 6% ($\Delta C \leq 6\%$). Gamma-ray coloration was carried out using a Co^{60} source and dose D = 10^{-4}-10^{-6} R; in this case ΔC reached 20%, and coloration was uniform. Changes in the Cr^{3+} concentration after coloration were determined to within 1% of the initial concentration C_0 using the intensity of the line due to the $+^1/_2 \to -^1/_2$ spin transition between the sublevels of the ground state $^4A_2$. All the ESR measurements were carried out by G. E. Arkhangel'skii.

Thermoluminescence

Heating of optically or γ-ray-colored crystals produced a single thermoluminescence peak at ~ 600°K. An analysis of the thermoluminescence curves by the general Antonov-Romanovskii method [41] made it possible to determine the parameters of electron color centers. The depth of the electron color centers and the frequency factor were the same for crystals colored by optical and γ-ray methods and their values were $\Delta E = 1.47 \pm 0.02$ eV and $w_0 = 10^{11}$ sec^{-1}; the kinetics of the process was of second order, and it was accompanied by quenching.

During heating the thermoluminescence spectrum was scanned repeatedly in the 400-820 nm range. This characteristic was of primary importance in the identification of the recombination centers. Since there was no agreement in the published literature,* the thermoluminescence spectrum was measured very carefully. In both cases the spectrum was identical with the photoluminescence spectrum of ruby measured under the same conditions [38].

These results demonstrated the existence of the dominant electron color centers and the dominant recombination centers (hole color centers), and the latter were the Cr^{4+} ions. Formation of the Cr^{4+} ions in colored ruby was deduced also from the ESR spectra ($\Delta M = 2$, g = 1.90, H $\parallel$ C_3). Observation of a forbidden paramagnetic transition did not by itself indicate a high concentration of the Cr^{4+} ions. The concentration of the color centers in ruby was at least 10^{18} cm^{-3} ($\Delta C \approx 20\%$ for $C_0 \approx 2 \cdot 10^{19}$ cm^{-3}) so that under these conditions the electron color centers could only be the Cr^{2+} ions. Formation of the Cr^{2+} ions as a result of coloration of ruby was confirmed by a reduction in the spin-lattice relaxation time of the Cr^{3+} ions [32]. Thus, irrespective of the coloration method, the dominant color centers in ruby were the Cr^{2+} and Cr^{4+} ions, i.e., the Cr^{3+} activator ions were capable of localizing an electron or a hole.

Additional Absorption and Quantum Efficiency Spectra

The additional absorption spectra of optically and γ-ray-colored crystals ($\Delta\varkappa_{opt}$ and $\Delta\varkappa_\gamma$) were plotted subject to a correction for the change in the Cr^{3+} concentration, and they were complex superpositions of the absorption bands in the spectral range 200-630 nm (Fig. 12). In the 300-630-nm range the $\Delta\varkappa_{opt}$ and $\Delta\varkappa_\gamma$ spectra did not differ greatly from crystal to crystal, whereas in the $\lambda <$ 300-nm range they were not identical for all crystals. It was established that the difference between $\Delta\varkappa_\gamma$ and $\Delta\varkappa_{opt}$ was due to the anomalously high concentration of titanium (~10^{18} cm^{-3}), and the charging of this element differed with the coloration method.

---

*Several investigators reported visual observations of the thermoluminescence but only one paper [31] gave thermoluminescence spectrum representing a symmetric wide band with λ_{max} = 640 nm.

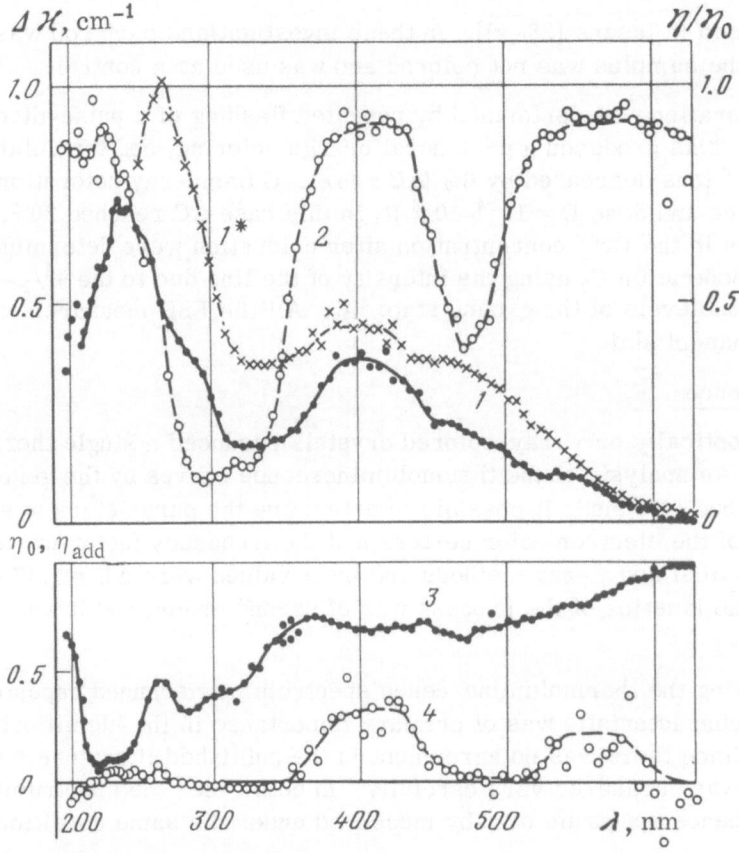

Fig. 12. Spectra of colored ruby: 1) additional absorption of optically colored ruby; 1*) γ-ray-colored ruby; 2) η/η_0; 3) η_0; 4) η_{add}. The ordinate of curve 1* was reduced threefold.

The quantum efficiency of colored ruby decreased selectively throughout the investigated spectral range (curve 2 in Fig. 12). A strong fall in η was observed in the region of weak activator absorption, and it indicated that the additional absorption was either completely inactive in respect of the R-line luminescence or was less active than the activator absorption. The experimental data on η_0, $\varkappa_0$, η, and $\varkappa$ (absorption coefficients and quantum efficiencies of colored and uncolored crystals, respectively), and $\Delta\varkappa = \varkappa - \varkappa_0'$ $\left(\varkappa_0' = \left(1 - \frac{\Delta C}{100}\right)\varkappa_0\right)$, were used to find the luminescence in the R-line region due to the additional absorption (curve 4 in Fig. 12):

$$\eta_{add} = \eta_0\left[\frac{\eta}{\eta_0}\left(1 + \frac{\varkappa_0'}{\Delta\varkappa}\right) - \frac{\varkappa_0'}{\Delta\varkappa}\right]. \tag{6}$$

The spectrum of η_{add} indicated the existence of two additional absorption regions with $\lambda_{max} \sim 550$ and 400 nm which were active in the excitation of the ordinary luminescence of ruby. This activity could be explained either by the sensitization of the luminescence of the Cr^{3+} ions as a result of a resonant transfer of the excitation energy from the color centers to the Cr^{3+} ions or by the optical annihilation of the color centers accompanied by the recombination luminescence:

$$Cr^{4+} + e \rightarrow Cr^{3+} + h\nu.$$

The last of these processes should give rise to optical bleaching of colored ruby, which was observed experimentally.

Optical Bleaching of Colored Ruby [40]

Bleaching of some parts of γ-ray-colored crystals was produced by illumination with mercury lamp radiation passed through a monochromator in such a way that the exposure was $\sim 10^{18}$-10^{19} photons. During illumination we recorded continuously the transmission at the bleaching-light wavelength (λ_{bl}) and periodically the absorption spectrum in the 200-630 nm range. It was found that illumination with the wavelengths $\lambda_{bl} = 549$ nm (10^{19} photons), 379.5, and 334 nm (10^{18} photons) bleached the crystals by about 80-90%, whereas illumination with $\lambda_{bl} = 302$ nm (10^{18} photons) reduced $\Delta\varkappa$ by 10-20%; $\lambda_{bl} = 470$ nm (10^{18} photons), 283 and 261 nm (10^{17} photons) caused no bleaching; illumination with $\lambda_{bl} = 160$-200 nm (produced by a DVS-200 hydrogen lamp, 10^{18} photons) generated new color centers.

An approximately uniform and strong reduction of the additional absorption throughout the investigated spectrum as a result of bleaching in the 550 and 400 nm bands indicated that: 1) the 550 and 400 nm bands were due to electron color centers of the same type; 2) these color centers predominated; 3) the optical annihilation of these electron color centers (i.e., of the Cr^{2+} ions) destroyed the dominant color centers (Cr^{4+} ions).

Thus, the inactive absorption band at 470 nm and the 330-200 nm region should correspond to the intracenter absorption in the Cr^{4+} ions. An increase in $C(Cr^{3+})$ was indeed observed as a result of optical bleaching, i.e., as a result of reduction in the additional absorption. Moreover, the value of $\Delta\varkappa$ measured at $\lambda = 470$ nm was quite accurately proportional to ΔC, and this experimental observation was used in the determination and allowance for small changes in ΔC as a result of optical bleaching in those cases when the precision of the determination of ΔC by the ESR method was insufficient.* The value of ΔC (in percent) for ruby with $C(Cr^{3+}) = 0.05$ at. % was deduced from the formula

$$\Delta C (\%) \approx 6.1 (\varkappa - 0.1), \tag{7}$$

where $\varkappa$ (cm^{-1}) was the absorption coefficient in the $E \perp C_3$ configuration.

Role of Titanium Impurities

All the investigated colored ruby crystals could be divided into two groups differing in respect of the spectra of $\Delta\varkappa$, η/η_0, and η_{add} in the 200-350 nm region. This difference was due to the anomalously high concentration of the titanium impurity ($\sim 10^{18}$ cm^{-3}). Titanium was deduced from the characteristic absorption band $\lambda_{max} \approx 220$ nm, which overlapped strongly the K absorption band of Cr^{3+}; it was also deduced from the ESR spectra in which Ti^{3+} gave rise to a wide band with g = 1.067 (H $\parallel$ C$_3$) at 4.2°K [42]. Crystals colored by γ irradiation and containing titanium exhibited an additional absorption band at $\lambda_{max} \approx 315$ nm, which was due to the electron color center. Bleaching in the 550-nm region enhanced the 315 nm band (Fig. 13), whereas in other parts of the spectrum the additional absorption decreased. Illumination with $\lambda_{bl} = 315$ nm reduced the $\Delta\varkappa$ spectrum uniformly throughout the investigated range, and the same effect was produced also by $\lambda_{bl} = 379.5$ nm because these two bands, associated with electron color centers of different origin, overlapped strongly.

Coloration of ruby containing titanium resulted in a strong reduction in the 220 nm band attributed to Ti^{3+}, and it gave rise to additional absorption in the 200-350 nm range. Clearly, coloration also altered the valence of titanium, and this effect was stronger than charging of

---

* The use of the 470 nm band for monitoring the relative change in ΔC was successful because uncolored ruby was practically transparent in this part of the spectrum and the value of $\Delta\varkappa$ could be measured accurately.

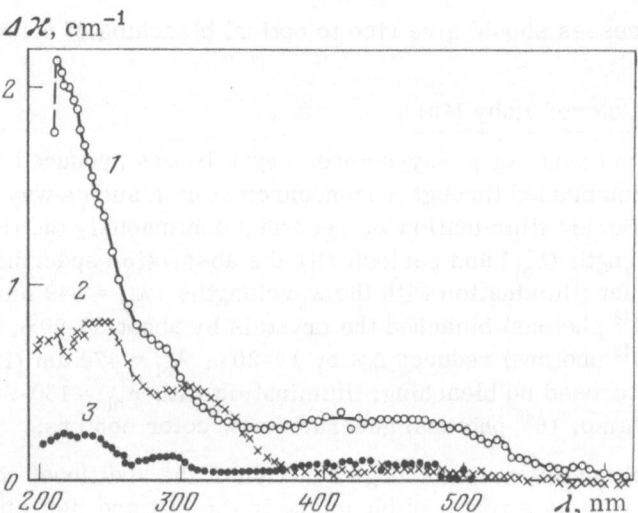

Fig. 13. Optical bleaching of γ-ray-colored ruby: 1) $\Delta\varkappa_\gamma$; 2) $\Delta\varkappa$ after bleaching with $\lambda_{exc} = 546$ nm light; 3) $\Delta\varkappa$ after second bleaching with $\lambda_{exc} = 315$ nm light.

chromium because a dip was always observed in the $\Delta\varkappa$ spectrum near 220 nm, which indicated that the correction $\Delta C(Ti^{3+})$, assumed to be equal to $\Delta C(Cr^{3+})$, was underestimated. Moreover, normalization to the 470 nm band of the $\Delta\varkappa_\gamma$ and $\Delta\varkappa_{opt}$ spectra of ruby with titanium indicated that in the 200-350 nm range we had $\Delta\varkappa_\gamma > \Delta\varkappa_{opt}$, and this region was less active in the γ-ray-colored ruby. This difference could be explained by the fact that Ti^{4+} centers were formed as a result of γ-ray irradiation, whereas they were not formed in the optical coloration process, i.e., in the latter case the Ti^{3+} ions were effective electron traps.

When the γ-ray-colored ruby crystals with titanium were excited in the 200-315 nm range, they emitted blue luminescence (curve 3 in Fig. 14) which was evidently associated with Ti^{3+}. The excitation of the blue luminescence in the 200-230 nm range could be attributed

Fig. 14. Spectra of ruby with titanium: 1) absorption spectrum of uncolored ruby; 2) Ti^{3+} impurity absorption band; 3) excitation spectrum of blue luminescence of titanium in γ-ray-colored crystal.

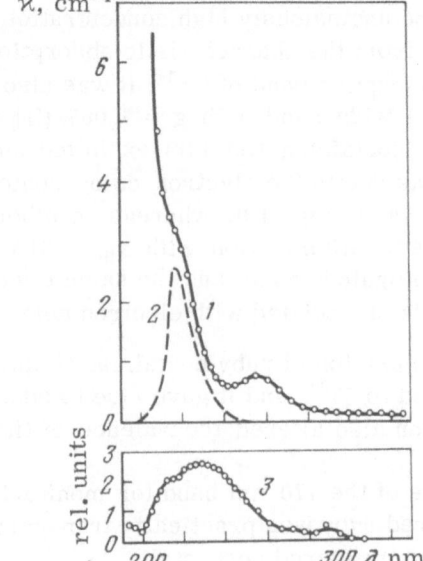

to the intracenter excitation of Ti^{3+}, but the excitation in the 230-315 nm range could only be due to the annihilation of electron color centers. The Cr^{2+} ions did not absorb in this part of the spectrum because the absorption in "pure" ruby was inactive. Consequently, in the 200-230 nm range there should be an absorption band due to the Ti^{2+} centers whose annihilation was accompanied by the blue luminescence of Ti^{3+} due to the recombination of electrons at Ti^{4+}:

$$(Ti^{2+}, Ti^{4+}) + h\nu \rightarrow (Ti^{3+}, Ti^{4+}) + e \rightarrow 2Ti^{3+} + h\nu.$$

The presence of titanium in ruby enhanced the coloration of the crystal by high-power optical illumination and produced more stable color centers (compared with Cr^{2+}) which were destroyed only by heating to $\sim 800°C$. However, even after many hours of annealing at 800°C the additional absorption disappeared only partly. Consequently, laser elements containing titanium would be expected to have a shorter service life, and the active characteristics of laser with such elements would be less stable.

§ 2. Identification of Additional Absorption Bands

Our identification consisted of separation of several regions (nominal bands) with strongly differing optical properties and attribution of these bands to a definite type of color center. In the band identification the main criteria were the band activity (or inactivity), behavior of the band under optical bleaching conditions, and differences between the $\Delta\varkappa$ spectra for crystals with and without titanium colored by different methods.

The active bands at 550 and 400 nm, associated with electron color centers of one type and exhibited by all the colored samples, were attributed to the Cr^{2+} centers. The inactive band at 470 nm and also the inactive absorption region 200-330 nm were attributed to the dominant hole centers, which were the Cr^{4+} ions.

The additional absorption band at 315 nm, which was of electron type in crystals with titanium, was attributed to the electron Ti^{2+} centers. Optically colored crystals containing titanium exhibited also active additional absorption bands at 246 and 220 nm. The activity of the 246 nm band could be explained by assuming that it was also due to the absorption by the Ti^{2+} ions. This was confirmed by the excitation of the blue luminescence associated with titanium. The activity of the 220 nm band was apparent and it was associated with the more effective disappearance of the Ti^{3+} band compared with the K band of Cr^{3+}, because these two bands competed in this part of the spectrum.

The absorption spectrum of the Ti^{4+} ions was clearly located in the 240-300 nm range, i.e., it overlapped completely the absorption by the Cr^{4+} ions because in γ-ray-colored crystals the absorption in this region was stronger and the 246 nm band was inactive. The Ti^{4+} ion had a filled 3p shell and a vacant 3d shell. Consequently, intraconfiguration transitions in Ti^{4+} were impossible, and one could expect only allowed interconfiguration transitions of the $3p \rightarrow 3d$ type. If the absorption in the 240-300 nm region was due to allowed transitions in the Ti^{4+} ions, the relatively low intensity of this absorption (~ 1 cm^{-1}) indicated that the concentration of the Ti^{4+} ions was very low: 10^{15}-10^{16} cm^{-3}.

Approximate positions of the absorption bands of the main color centers in ruby were as follows: 550 and 400 nm for the Cr^{2+} centers, 470 and 330-200 nm (two bands) for Cr^{4+}, 315 and 246 nm for the Ti^{2+}, and 300-240 nm for Ti^{4+}. It would be interesting to compare these results with theoretical predictions of the splitting of the ground-state terms ^3F and ^5D (electron configurations $3d^2$ and $3d^4$) in a field of C_{3v} symmetry.

In the case of the isoelectronic ions Cr^{4+} and Ti^{2+} ($3d^2$ configuration) the expected splitting should produce three wide absorption bands [43], and the ground-state term of the Cr^{2+} ion ($3d^4$ configuration) should split into two (for Dq < 2B) or three (for Dq > 2B) levels. In the former case we should observe one wide absorption band, and in the latter case there should be two bands. Thus, the experimental results agree with the theoretical predictions in respect of the number of bands attributed to each type of center. However, there is a disagreement in respect of the nature of the 315 nm band, which is observed for "rust-colored" rubies unsuitable for use in lasers. In [17] this band is attributed to the Cr^{4+} ions because it appears in crystals grown under oxidizing conditions, which are favorable for the formation of the Cr^{4+} ions. Following the conclusions reached in [17], we calculated the spectrum of Cr^{4+} in Al_2O_3 and found the positions of the absorption bands [44]:

$$\text{for} \quad a^3T_1 \to {}^3T_2 \quad \lambda_{max} \simeq 552 \text{ nm},$$
$$\text{for} \quad a^3T_1 \to b^3T_1 \quad \lambda_{max} \simeq 317 \text{ nm},$$
$$\text{for} \quad a^3T_1 \to {}^3A_2 \quad \lambda_{max} \simeq 263 \text{ nm}.$$

These positions are not in agreement with our experimental data indicating that the 315 nm band is associated with electron color centers: probably with Ti^{2+} but not with Cr^{2+} or Cr^{4+}.

The 470-nm band is exhibited by ruby crystals grown under oxidizing and reducing conditions. Clearly, this means that one cannot grow ruby crystals with Cr^{2+} free of Cr^{4+} and vice versa because the Cr^{3+} ions are capable of compensating positive and negative excess charge by accepting or giving up one electron. In the presence of Ti^{3+}, which is more effective as an electron trap than Cr^{3+}, the conditions are favorable for the formation of Cr^{4+} and then stable Ti^{2+} centers (315 nm band) and Cr^{4+} (470-nm band and 200–330-nm region) are observed, and these are the centers responsible for the rust color of ruby. This color cannot be removed completely by annealing in air because cooling causes free electrons to be captured primarily by the Ti^{3+} ions. A more effective method may be optical bleaching with $\lambda_{bl} = 315$ nm. However, reduced crystals should be colored quite rapidly by high-power optical radiation.

The results obtained make it possible to construct an energy-level system of the Cr^{2+} ion in the forbidden band of corundum. The ground state of Cr^{2+}, deduced from the thermoluminescence spectra, is located 1.5 eV below the conduction band. Excited levels, corresponding to the additional absorption bands at ~ 550 and ~ 400 nm, are located 2.2 and 3.1 eV higher than the ground state, i.e., they are localized above the bottom of the conduction band, but because of the bending of this band in the k space they are still in the forbidden gap. This situation is in qualitative agreement with the observation that the photoionization of Ce^{2+} by indirect optical transitions or tunneling from the excited state is an order of magnitude more effective in the 440 nm band than in the 550 nm band.

§3. Influence of Color Centers on
Stimulated Emission Threshold of a Ruby
Laser [45]

When a ruby laser element is used for a long time, its stimulated emission threshold gradually rises. Moreover, this threshold rises when the active element is bombarded with γ rays in a dose $D_e > 2 \cdot 10^4$ R [46]. When the γ-ray does not exceed $4 \cdot 10^3$ R, the output energy increases for a given pumping rate, and this has been attributed to "optically stimulated thermoluminescence." If this term means the pump-like dissociation of the electron trapping levels formed by γ-ray irradiation or filled as a result of this irradiation, it is difficult to see how such small doses could account for the threefold increase in the stimulated radiation energy. A dose $D_e \sim 10^3$ R can produce 10^{16}–10^{17} cm^{-3} electron color centers, and, consequently, even when all these centers dissociate during one pump pulse, the addition of 10^{16}–10^{17} cm^{-3} photons cannot contribute significantly to the output energy. The actual contribution will be much less because of the competing processes. A phenomenological analysis of these phenomena in [46, 47] has ignored the physical causes responsible for the rise of the threshold and output energy and has not resulted in any practical conclusions.

The change in the laser characteristics of ruby crystals is undoubtedly due to the formation of color centers, which give rise to additional absorption in the 200-630 nm range, and this absorption reduces the efficiency of the optical pumping and increases the threshold. The rate of transitions in color centers is approximately an order of magnitude higher than the rate of transitions in the Cr^{3+} ions, and, therefore, even a small number of such color centers ($\sim 10^{18}$ cm^{-3}) can produce an inactive absorption comparable with the activator absorption and the former reduces the luminescence quantum efficiency of the Cr^{3+} ions.

The relationship between the formation of color centers and the rise of the stimulated emission threshold was studied in the laser rods and composite laser elements placed in a resonator with external dielectric mirrors. The pump radiation was generated in a two-lamp close-fitting enclosure. A series of pump pulses was applied to the samples, and the $\varkappa(\lambda)$ spectra were determined in the intervals between the pump pulses. The efficiency of the laser enclosure was calibrated using a standard crystal. Moreover, allowance was made for the temperature dependence of the threshold.

The dependence of the stimulated emission threshold on the total electrical energy supplied in the form of pump pulses was determined (Fig. 15). This dependence was obtained for pump energies close to the threshold value. The trapezoidal shape of the pump light pulses, with a leading edge of $\sim 0.15 \cdot 10^{-3}$ sec and a total duration of $\sim 1 \cdot 10^{-3}$ sec, ensured that the threshold energy was measured correctly when the pump power was increased. The initial part of the dependence was nearly linear but at higher pump energies the curve reached saturation. This saturation could be attributed to the dynamic equilibrium between the formation of color centers and their optical dissociation. A parallel check of the $\varkappa(\lambda)$ and $\eta(\lambda)$ spectra showed that changes in these spectra reached saturation at the same time as the threshold energy. Prolonged optical pumping increased the absorption throughout the investigated range 200-625 nm and reduced the quantum efficiency of the R luminescence in the range 230-625 nm. The concentration of the Cr^{3+} ions, measured by the ESR method after reaching saturation, decreased by $6 \pm 0.5\%$.

The appearance of inactive absorption by color centers and the associated reduction in the luminescence efficiency increased the excitation rate needed to achieve a population inversion of the ^2E state. The inactive absorption due to color centers played the role of a passive filter in respect of the pump radiation, and it was clearly the main cause of the rise of the stimulated emission threshold. The other factor, also associated with the formation of color centers, was the influence of the spectrum of the Cr^{3+} ions, which was manifested by a reduction in the quantum efficiency in the case of resonance excitation of γ-ray-irradiated and optically colored crystals (Fig. 7). The observed fall in η (R) could not be attributed to the inactive additional absorption in the region of the luminescence spectrum, and one had to assume that the excitation was transferred from Cr^{3+} to color centers or that the probabilities of some of the transitions in the Cr^{3+} ions changed in the presence of color centers. The latter factor could be related directly to the rise in the output energy which occurred under certain pumping conditions.

The results plotted in Fig. 15 were influenced by the measurement method (pumping near the threshold). At higher pumping rates one could expect a steeper rise of the threshold energy and a higher saturation plateau, but it would then be difficult to measure exactly the threshold without reducing the pumping rate and such a reduction would have disturbed the dynamic equilibrium between the formation and dissociation of color centers so that the crystal could become partly bleached.

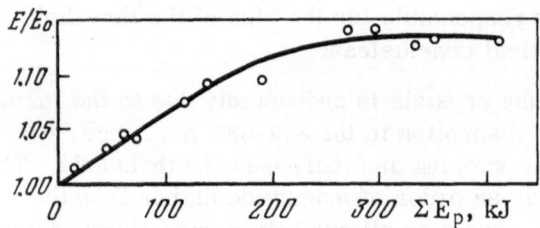

Fig. 15. Dependence of the stimulated emission threshold of a ruby laser on the total pump energy.

Thus, we may regard as established that there is the relationship between the formation of color centers as a result of high-power optical pumping and an increase in the stimulated emission threshold. The multistage photoionization of the Cr^{3+} ions makes it impossible to prevent completely the formation of color centers in an active laser element and to stabilize the threshold. However, the rate of formation of these centers and their equilibrium concentration can be reduced considerably if the $\lambda < 350$ nm range is excluded from the pump radiation because it is this range that is most effective in the formation of color centers and least effective in pumping a ruby laser.

About 14% of the optical energy of a pulse-discharge lamp is represented by the wavelength range $\lambda < 350$ nm. Therefore, we considered the possibility of an active filter transforming this ultraviolet part of the pump spectrum into radiation corresponding to the absorption bands of ruby. Simple estimates indicated that in the ideal case (an activated quartz tube emitting luminescence in the 400 nm range) one could increase the quantum efficiency of the pump radiation by 11%. We built and tested tubes made of different luminescent glasses which were used to enclose a ruby element.* Unfortunately, the nonoptimal filter geometry, which resulted in considerable reflection of the pump radiation and inefficient extraction of the converted light, failed to produce the expected increase in the pump efficiency; the stimulated emission threshold actually increased by 6-15%. The best result, which simply maintained the earlier threshold, was achieved by the use of organic films containing dyes and deposited directly on the active element. However, such organic films were rapidly damaged. This failure to increase the efficiency of pump radiation by transforming ultraviolet should not be regarded as final proof that filters cannot be used for reducing the concentration of color centers and stabilizing the active laser characteristics, because this may be achieved by the use of high-quality passive filters.

§4. Influence of Color Centers on the Cr^{3+}

Ion Spectrum [48]

It follows from the preceding discussion that the lifting of the forbiddenness on transitions between states of the same parity and multiplicity occurs because of the mixing of the states of different parity by the odd component of the crystal field. The intensities of such forbidden transitions are described by

$$\sigma \simeq \sigma_{all} \left| \frac{(V_{odd})_{ad}}{E_a - E_d} \right|^2, \qquad (8)$$

where $(V_{odd})_{ad}$ is the matrix element of the odd component of the crystal field, $E_a - E_d$ is the gap between the nearest levels of different parity ($\approx 10^5$ cm^{-1} in ruby), and σ_{all} is the intensity

_____

* The author is deeply grateful to N. A. Gorbacheva for her great help in the synthesis of luminescent glasses for such active filters.

of an allowed electric-dipole transition. Clearly, the intensity of the forbidden transitions is sensitive to a change in V_{odd}.

Several experimental observations suggest that the Cr^{3+} ions interact with color centers. This interaction is manifested by a reduction in the relaxation time of spin transitions in the Cr^{3+} ion and by a reduction in η (R) (see curves 4 and 6 in Fig. 7), which cannot be attributed to the inactive absorption in color centers. The temperature dependence of η (R) obtained for colored ruby crystals indicates the appearance of a nonradiative channel of excitation relaxation, whose probability depends weakly on temperature in the range 200-300°K. The rise of the probability of nonradiative decay of the ^2E state may be attributed to the action of local Coulomb fields, due to excess charges $\pm e$ of the color centers which reduce the symmetry of the Cr^{3+} position. Consequently, V_{odd} increases and the probability of forbidden transitions becomes greater. Clearly, for the same reason the probability of the $d-d$ transitions in the Cr^{2+} and Cr^{4+} ions is almost an order of magnitude higher than the probability of transitions in the Cr^{3+} ion. The random distribution of excess charges in a colored crystal should give rise to an additional inhomogeneous Coulomb field which may cause inhomogeneous broadening of the narrow spectral lines.

Broadening of the ESR lines corresponding to the $^3/_2 \to {}^1/_2$ and $-^3/_2 \to -^1/_2$ transitions between the ground-state levels of the Cr^{3+} ion has been observed in ruby crystals bombarded with x rays [32]. These spin transitions occur between levels belonging to different Kramers doublets. An inhomogeneous Coulomb field due to color centers results in a scatter of the initial splitting constant $|2D|$, which is manifested as broadening of the ESR lines. The application of an external electric field $\mathscr{E} \parallel C_3$ splits the $^3/_2 \to {}^1/_2$, $-^3/_2 \to -^1/_2$ lines in the ESR spectrum [49] and the splitting is proportional to $\mathscr{E}$, the coefficient of proportionality being

$$\frac{\partial H}{\partial \mathscr{E}} = (0.93 \pm 0.04) \cdot 10^4 \ \text{G} \cdot \text{V}^{-1} \cdot \text{cm}. \qquad (9)$$

Our γ-ray-colored ruby crystals also exhibited broadening of the $^3/_2 \to {}^1/_2$ (H $\parallel C_3$) line. This broadening was used to estimate the average field of color centers, which was $2 \cdot 10^4$ and $2.8 \cdot 10^4$ V/cm in samples with $C(Cr^{3+}) = 0.01$ and 0.03 at. %, respectively, colored under identical conditions with a dose $D_e \approx 10^6$ R. Since these experimental observations demonstrated a considerable influence of the additional Coulomb field generated by color centers, we investigated the intensities of the intercombination transitions $^4A_2 \to {}^2E$, $^2T_2$ at 77 and 300°K in γ-ray-colored crystals ($D_e \approx 10^6$ R).

Rise of the Intensity of Intercombination Transitions

in Colored Ruby

An investigation of the intensities of the $^4A_2 \to {}^2E$, $^2T_2$ transitions was carried out for two polarizations of the incident light using instruments capable of spectral resolution of 1.1 and 0.7 cm^{-1} in the R- and B-line regions, respectively. The width of the absorption bands in the R-line region of an uncolored crystal (77°K) was 1.8-2 cm^{-1} and it did not decrease when the spectral width of the slit was reduced to 0.4 cm^{-1}. Consequently, the investigated crystals were of poor quality and the apparatus-induced distortions of the spectra recorded at 77°K were relatively slight.

The concentration $C(Cr^{3+})$ in colored crystals decreased by 18.5% but, in spite of this reduction in $C(Cr^{3+})$ the absorption spectrum in the R-line region was practically unaffected at 300°K, whereas at 77°K the absorption even increased. There was no doubt that $C(Cr^{3+})$ decreased because this was deduced from the intensity of the $+^1/_2 \to -^1/_2$ spin transition between the Kramers doublet sublevels which were not affected by the Coulomb fields of the

TABLE 4. Increase in Oscillator Strength of
Intercombination Transitions in Colored Ruby

Absorption band	Electron transition	f_γ/f_0		T, °K
		$E \perp C_3$	$E \parallel C_3$	
$R_1 + R_2$	$^4A_2 \rightarrow {}^2E$	1,21	1,30	300
		1,43	1,58	77
$B_1 + B_2$		1,29	1,39	300
	$^4A_2 \rightarrow {}^2T_2$	1,26	1,35	77
B_3		2,06	1,63	300
		1,27	1,45	77

color centers. Consequently, the presence of color centers increased the oscillator strength of the $^4A_2 \rightarrow {}^2E$ transition. The rise of the absorption after γ-ray coloration was also observed in the B-line region ($^4A_2 \rightarrow {}^2T_2$), where there were practically no spectral distortions.

Table 4 lists the qualitative changes in the intensities of the absorption in a γ-ray-colored crystal corresponding to a change in the Cr^{3+} concentration amounting to $\Delta C = 18.5\%$. The relative change in the oscillator strength of the transition in question was determined from the ratio of the integrals of the corresponding absorption bands in γ-ray-colored (f_γ) and uncolored (f_0) crystals allowing for the change in the concentration of the Cr^{3+} ions:

$$\frac{f_\gamma}{f_0} = \frac{\int \varkappa_\gamma (\nu)\, d\nu}{(1 - \Delta C/100) \int \varkappa_0 (\nu)\, d\nu}. \tag{10}$$

It is clear from Table 4 that the value of f_γ increased considerably in all cases. The greatest possible error, due to spectral distortions, was expected for the value of f_γ/f_0 in the case of the $^4A_2 \rightarrow {}^2E$ transition in the $E \perp C_3$ configuration at 77°K. However, there was no doubt that the transition increased because f_γ increased more strongly in the $E \parallel C_3$ polarization, i.e., under the conditions such that the spectral distortion should have less effect on the absorption integral. The precision of the determination of the absorption integrals in the R-line region at 300°K and in the B-line region at both temperatures was at least 10%.

We investigated also the dependence of f_γ/f_0 in the R-line region on the value of ΔC at 77 and 300°K. The value of ΔC was altered (by changing the concentration of color centers) as a result of partial annealing of γ-ray-colored crystal; the value of ΔC was deduced from $\Delta \varkappa$ in the 470-nm band using Eq. (7), and it was checked three times using the ESR spectra. Ten experimental points were obtained in the range of ΔC from 0 to 18.5%, which represented an approximately linear increase in f_γ/f_0 with rising ΔC. It should also be noted that in the case of a γ-ray-colored crystal at 77°K the absorption lines broadened slightly (~ 0.2 cm^{-1}), and there was a small opposite shift of the R_1 and R_2 lines (again within 0.2 cm^{-1}).*

Reduction in the Metastable-State Lifetime

An increase in the probability of the $^4A_2 \rightarrow {}^2E$ transition should also give rise to a proportional reduction in the metastable-state lifetime in a colored crystal, i.e., $f_\gamma/f_0 = \tau_0/\tau_\gamma$. However, a reduction in the concentration of the Cr^{3+} ions by 20% as a result of coloration should increase τ by $\sim 10\%$ because of weakening of the exchange interaction [50]. Thus, simultaneous reduction of these two approximately equal but opposite effects should largely neutralize the influence of coloration on τ. The values of τ_0 and τ_γ at 300°K were determined

---

* The reference lines were 6929.47 Å (Ne) and 6907.16 Å (Hg).

(to within ± 5%) from the luminescence decay curves and no significant difference was found between them. A reduction in τ_γ was established only after the development of a differential method for measuring τ_γ/τ_0, which revealed the difference between τ_γ and τ_0 amounting to 0.2% [51]. In the case of a γ-ray-colored crystal with $\Delta C = 20\%$, the value of τ_γ was found to be $2 \pm 0.5\%$ less than τ_0, but after partial thermal bleaching and reduction in ΔC to 14%, the difference between τ_γ and τ_0 also decreased to $1.5 \pm 0.3\%$. Finally, after complete annealing and recovery of the initial value of $C(Cr^{3+})$, it was found that the decay time of the ^2E state in an annealed crystal agreed to within 0.2% with the value obtained for the control crystal. Other crystals with laser-type chromium concentrations also exhibited a reduction in τ_γ by 1–2.5% depending on the quality of the crystals and on the dose D_e.

An increase in the probability of intercombination transitions due to the formation of color centers was due to a reduction in the symmetry of the Cr^{3+} positions under the action of the Coulomb fields of the charged color centers. This effect should alter the laser character- istics of ruby crystals. An increase in the concentration of the color centers should increase also the inactive absorption, reducing the pumping efficiency, and raise the probability of res- onance transitions. The latter effect should increase the gain in the active medium and re- duce the specific losses. In this situation it should be possible to optimize the output laser energy at some pumping rate, as indeed observed in [46] for a low concentration of color cen- ters. In this sense, the presence of few color centers (in amounts not exceeding 10^{17} cm^{-1}) could have a positive effect on the laser operation but, because of the complex dynamics of the processes of formation and dissociation of color centers, it would be practically impossible to maintain a constant color concentration, and the change in this concentration during the operation of a laser would result in an instability of the laser characteristics.

In connection with the discovery of the increase of the probability of resonance transitions in the presence of color centers one should mention the incorrectness of the method of deter- mination of $C(Cr^{3+})$ in ruby from the intensity of absorption in the R-line region. In particular, monitoring of the changes in $C(Cr^{3+})$ during coloration of ruby by this method led Stickley et al. [36] to an incorrect conclusion that $C(Cr^{3+})$ was constant and this resulted in an incorrect iden- tification of the color centers.

CHAPTER IV

EFFECT OF AN EXTERNAL ELECTRIC FIELD ON THE SPECTRUM OF Cr^{3+} IN RUBY [52]

If, following [2], we assume that the intensities of the $^4A_2 \rightarrow {}^4T_2$ and $^4A_2 \rightarrow {}^2E$ transitions are related linearly by the spin–orbit interaction constant, an increase in the transition prob- ability in the presence of color centers should be regarded as a consequence of an increase in the intensity of absorption in the U band. A direct observation of the change in the intensity of the U band is very difficult because it is overlapped completely by the additional absorption due to the presence of color centers. However, this problem is of fundamental importance be- cause its solution would provide a check of the hypothesis of the dominant role of the intracen- ter spin–orbit interaction in the lifting of the forbiddenness of the intercombination transitions.

An increase in the probability of intercombination transitions in the presence of color centers may be explained by a partial violation of the parity rule because of an increase in the low-symmetry component of the crystal field due to the presence of local Coulomb fields of charged color centers. If this hypothesis is correct, we may expect a similar effect on appli- cation of an external electric field to uncolored ruby. Preliminary estimates of the average field due to color centers ($\sim 3 \cdot 10^4$ V/cm) have suggested that an increase in the probability of forbidden transitions in an external electric field should also be considerable. Thus, an investigation of this effect in its pure form would make it possible to discriminate between the influence of the change in the odd component of the crystal field on the spectrum of Cr^{3+} and other types of interactions which occur in the presence of color centers.

Since the initial and final states of the $^4A_2 \rightarrow \, ^2E, \, ^2T_2$ transitions are Kramers doublets, we cannot expect a significant normal Stark effect but only a Stark shift of the Cr^{3+} spectrum in an external field $\mathscr{E} \parallel C_3$, which can lift the double orientational degeneracy manifested at low temperatures in the "pseudo-Stark" splitting of the R and B lines [53-55]. In the absence of an external field the orientationally degenerate positions are energetically equivalent and differ only in the sign of $V_{odd}(0)$. The application of an axial field $\mathscr{E} \parallel C_3$ should enhance the odd component of the crystal field in one position and reduce it in the other position. If we assume a linear relationship between $V_{odd}(\mathscr{E})$ and the perturbing field $\mathscr{E} \parallel C_3$ at two inequivalent positions in the form $V_{odd}(\mathscr{E}) = V_{odd}(0) \pm \alpha\mathscr{E}$ and if we allow for the fact that the intensity of the forbidden transition is proportional to $|V_{odd}|^2$ [according to Eq. (8)], we find that an increase in the external field should result in an increase (quadratic in respect of $\mathscr{E}$) in the transition intensity:

$$\Delta f \propto |\alpha\mathscr{E}|^2, \tag{11}$$

where $\Delta f = f_{\mathscr{E}} - f_0$ is the increase in the absorption integral of both split components under the influence of the perturbing electric field. The pseudo-Stark split components should then be of different intensities. This conclusion follows directly from the ideas underlying the crystal field theory but, to the author's knowledge, no experimental investigations of the change in the intensities of forbidden transitions under the influence of an electric field have yet been made.

§1.   Absorption Intensities in the Presence

of an Electric Field

We investigated a ruby sample of $0.6 \times 8 \times 20$ mm dimensions with $C(Cr^{3+}) \approx 0.15$ at. % at 77°K using the apparatus shown in Fig. 6. Electrodes were deposited on the 8×20 mm ($\mathscr{E} \quad C_3$)

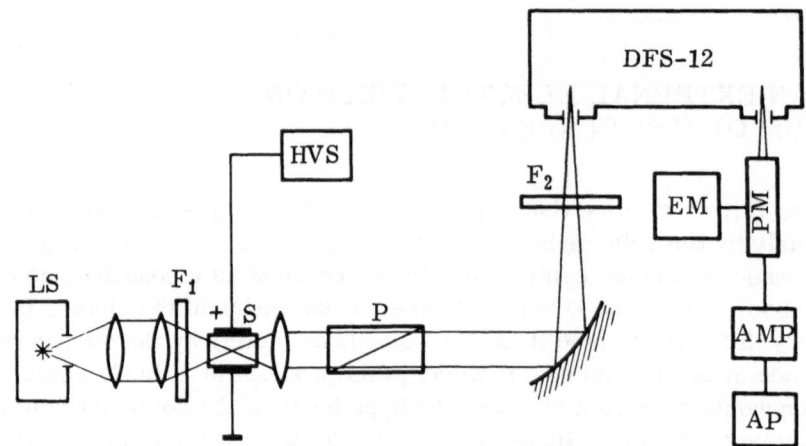

Fig. 16. Schematic diagram of the apparatus for investigating the behavior of a ruby crystal in an external electric field. Here, LS is a stabilized light source; $F_{1,2}$ are filters; S is a sample; P is a polarizer; HVS is a high-voltage source; DFS-12 is a spectrograph; EM is an electronic modulator; PM is a photomultiplier; AMP is a selective amplifier; AP is an automatic plotter.

surface. A light beam was directed on the 0.6 × 20 mm surface and it was perpendicular to C_3. We investigated the absorption spectra in the R- and B-line regions and also in the 5780-3680 Å region in two polarizations. The resolution in studies of narrow and wide bands was 0.2 and 0.5 Å, respectively. A sample with high Cr^{3+} concentration was selected because of the desirability to reduce as much as possible the spectral distortions (the half-width of the R absorption lines increased with the concentration C of Cr^{3+}).

An external electric field (up to $2.4 \cdot 10^5$ V/cm) was applied periodically: A crystal was subjected to a field for 5 sec, and then the field was switched off during the next 5 sec; this enabled us to record simultaneously two spectra and to avoid errors of the sign of the effect in the case of a weak influence of the field.

Wide-Band Spectra

The field effect in the wide U and Y absorption bands could be deduced from a change in $\varkappa$. The transmission spectrum was recorded in the 5780-3680 Å range by modulation with a field of 135 kV/cm using two polarizations of the exciting light (the absorption in the B-line region was ignored). This method made it possible to record reliably the minimum change in $\varkappa$ amounting to 0.02-0.03 cm^{-1}. The transmission showed no changes in the absorption, i.e., the intensities of the U and Y bands in a field of 135 kV/cm remained constant to within 2%.

Absorption in R-Line Region

Figure 17 shows the absorption spectra obtained in the R-line region for two polarizations using two values of the field $\mathscr{E}$ and in the absence of the field. It is clear from this figure that the absorption integral increased considerably on application of an external field. The changes in the absorption integral $f_{\mathscr{E}}/f_0$ are given in Table 5.

The field effect appeared more strongly for the $E \perp C_3$ polarization but in the case of the R_1 line it was strong in both polarizations. As expected, the absorption intensities of the split components were unequal: The short-wavelength components were stronger. Possible apparatus distortions of the spectra could increase the absorption integral when the lines were split in the $E \perp C_3$ case.

However, in the $E \parallel C_3$ case there should be no significant error in the value of $f_{\mathscr{E}}/f_0$ due to the spectral distortion because in this polarization the absorption was weak. The unavoidable apparatus distortions of the spectra yielded only a qualitative result indicating an increase in the intensity of the $^4A_2 \rightarrow {}^2E$ transition under the influence of a perturbing electric field.

Absorption in B-Line Region

Figure 18 shows changes in the B_1 and B_2 line spectra on application of an external field (the B_3 line was not affected by the field). These lines were approximately an order of magnitude wider than the R lines so that the apparatus distortions were unimportant. Table 6 gives

TABLE 5. Changes in Absorption Integral
in R-Line Region Due to Field $\mathscr{E} \parallel C_3$ (77°K)

$\mathscr{E}$, kV/cm	$E \perp C_3$		$E \parallel C_3$	
	R_1	R_2	R_1	R_2
0	1	1	1	1
67	1.20	1.33	1.10	1.08
135	1.85	1.51	1.21	1.06
157	1.85	1.57	1.26	1.17

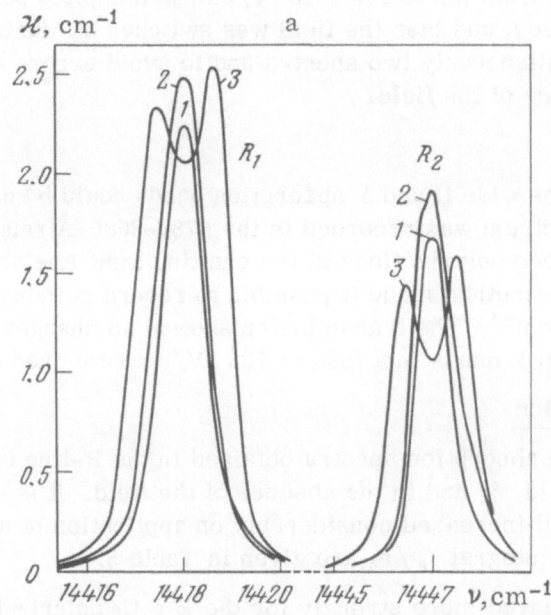

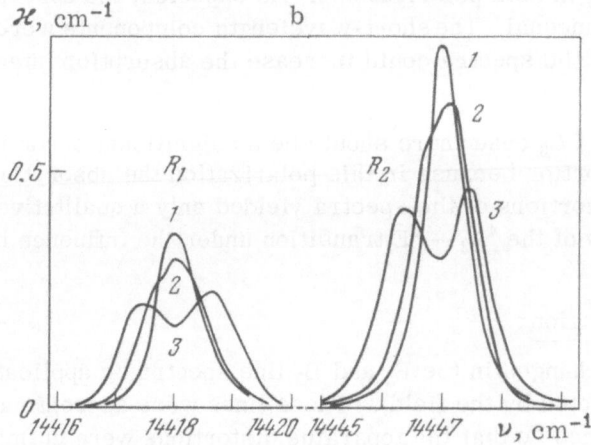

Fig. 17. Absorption spectra in the R-line re-
gion obtained in an external electric field $\mathscr{E} \parallel C_3$
at 77°K: a) $E \perp C_3$; b) $E \parallel C_3$; 1) $\mathscr{E} = 0$;
2) $\mathscr{E} = 67$ kV/cm; 3) $\bar{\mathscr{E}} = 160$ kV/cm.

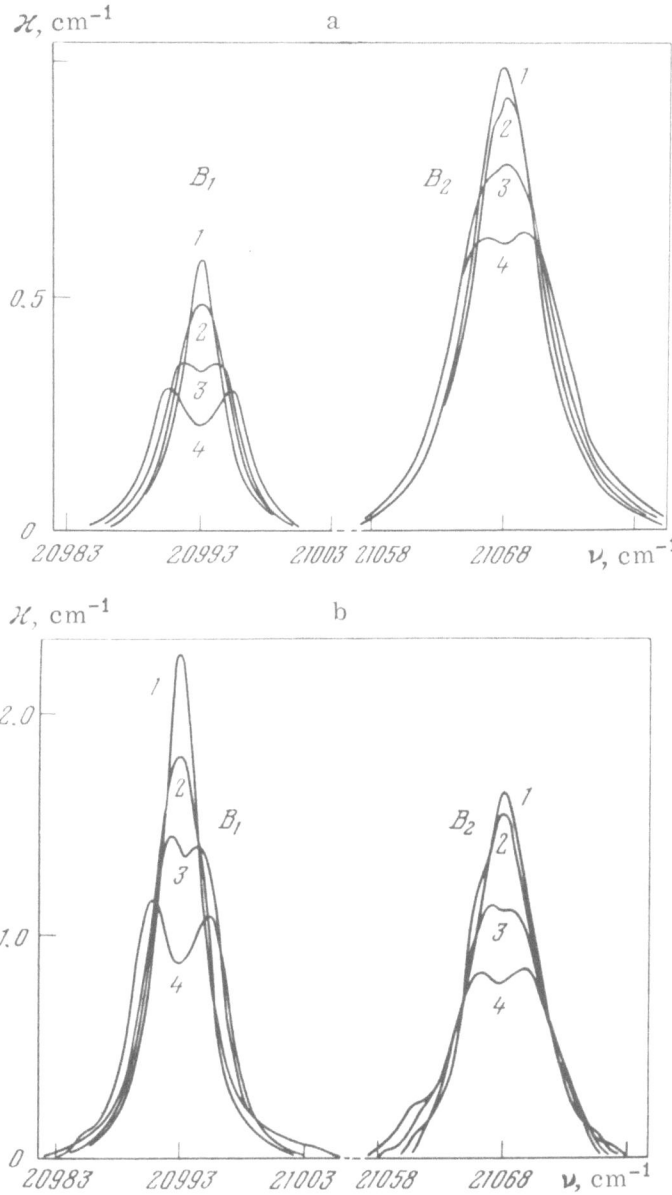

Fig. 18. Absorption spectra in the $B_{1,2}$-line region obtained in an external electric field $\mathscr{E} \parallel C_3$ at 77°K:
a) $E \perp C_3$; b) $E \parallel C_3$; 1) $\mathscr{E} = 0$; 2) $\mathscr{E} = 67$ kV/cm;
3) $\mathscr{E} = 140$ kV/cm; 4) $\mathscr{E} = 185$ kV/cm.

TABLE 6. Changes in Absorption Integral
in B-Line Region Due to Field $\mathscr{E} \parallel C_3$ (77°K)

$\mathscr{E}$, kV/cm	$E \perp C_3$		$E \parallel C_3$	
	B_1	B_2	B_1	B_2
0	1	1	1	1
33	0.97	1.00	1.00	0.99
67	1.05	0.97	0.98	0.97
133	1.02	1.04	1.03	0.88
185	1.01	0.99	0.94	0.71

the results of measurements of the absorption integrals carried out with an accuracy of 5%. Within the limits of the experimental error the intensities of the B lines remained constant with the exception of the B_2 line ($E \parallel C_3$), which became much weaker when the field intensity was increased.

Increase in the Probability of the $^2E \rightarrow {}^4A_2$ Radiative

Transition

An investigation of the luminescence spectra in the R-line region in the presence of an external field revealed not only a pseudo-Stark splitting of the R lines, but also a considerable increase in the intensities of these lines in both polarizations. The excitation corresponded to the U and Y bands, and, because the absorption in ruby in the excitation region was not affected by the field, the observed effect indicated an increase in the radiative efficiency of the R lines (i.e., an increase in the "weight" of the R lines in the luminescence spectrum).

The rise of the intensity of the R-line luminescence was undoubtedly of scientific interest, and it could also be of practical importance. We investigated the field dependences of the integrated intensities of each line in two polarizations. The field dependences were determined using the apparatus shown in Fig. 16; the spectral width of the entry slit was 0.4 cm^{-1} and the exit slit was 4.5 cm^{-1} with the λ tuned to the line center. Since the half-widths of the R_1 and R_2 luminescence lines, measured using a resolution of 0.4 cm^{-1}, were 1.45 and 1.25 cm^{-1} ($\mathscr{E} = 0$), we determined experimentally a quantity which was proportional to the integrated intensity of the luminescence of the whole line. Figure 19 shows the results of measurements of the field dependences of the integrated intensities of the R_1 and R_2 lines. When the field was increased, the relative luminescence intensity first rose superlinearly and then exhibited a tendency to saturation. As in the absorption case, the field effect was greater in the $E \perp C_3$ case but, in contrast to the absorption, the intensity of the R_2 line changed more. Similar results were also obtained when the excitation corresponded to the Y band.

The results obtained allowed us to estimate the radiative efficiency of the R lines. This estimate was obtained allowing for the contribution of each component in the σ and π spectra and for its relative weight in the total R-line luminescence; moreover, the polarizability of the luminescence by the monochromator was taken into account. The results obtained are plotted in Fig. 20, which shows an increase in the radiative efficiency reaching 30% for $\mathscr{E} = 225$ kV/cm.

This increase in the radiative efficiency of the R lines could not be explained by a trivial redistribution of the luminescence intensity between different parts of the spectrum as a result of the splitting of the R_1 and R_2 lines and the unavoidable reduction in the reabsorption of the luminescence. This reduction in the reabsorption because of the pseudo-Stark splitting could distort the results obtained in fields $\mathscr{E} < 150$ kV/cm until the splitting components were fully separate. It is clear from Figs. 19 and 20 that the field effect in $\mathscr{E} = 150$ kV/cm was

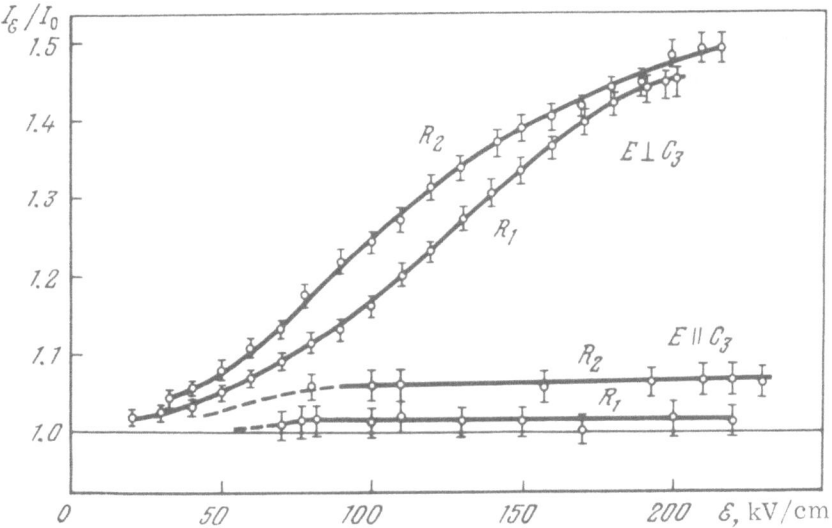

Fig. 19. Dependences of the intensities of the R luminescence lines of ruby on the applied electric field $\mathscr{E} \parallel C_3$ at 77°K for $\lambda_{exc} =$ 560 nm.

far from saturation and, moreover, this effect altered more strongly the intensity of the R_2 line which was the weaker component in the σ spectrum. Consequently, the rise of the radiative efficiency indicated an increase in the luminescence quantum efficiency in an external field $\mathscr{E} \parallel C_3$ when the excitation was provided in the U and Y bands. The increase in η (U, Y) could be due to an increase in the probability of the $^4T_2$, $a^4T_1 \rightarrow {}^2E$ nonradiative transition, increase in the probability of the $^2E \rightarrow {}^4A_2$ transitions, or reduction in the probability of the $^4T_2$, $^4T_1 \rightarrow {}^4A_2$ nonradiative transitions. However, the field dependence of the radiative efficiency in the excitation in the U- and Y-band regions and the experimentally observed increase in the probability of the $^4A_2 \rightarrow {}^2E$ transition suggested that the perturbation by the $\mathscr{E} \parallel C_3$ field increased the probability of the radiative decay of the ^2E state.

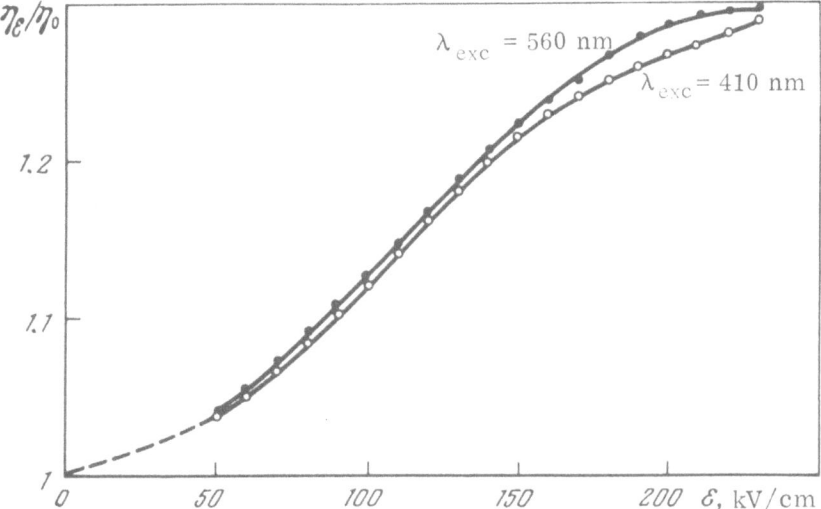

Fig. 20. Dependences of the radiative efficiency of the R lines on the electric field $\mathscr{E} \parallel C_3$ at 77°K.

It was reported in [56] that η (U) and η (Y) changed by not more than 5% between 77 and 300°K. Hence, an increase in the radiative efficiency of the R lines by 30% in a field $\mathscr{E} > 200$ kV/cm was due to an increase in η (R) by 10-30% (20% was the maximum increase in the radiative efficiency due to reduction in the reabsorption). The value of η (R) at 77°K was found to be 0.79 ($\mathscr{E} = 0$), i.e., an increase in η (R) by 10-30% in an external field was, in principle, possible.

§ 2. Kinetics of Luminescence Decay in an External Electric Field

The same sample was used in an investigation of the luminescence decay in the R-line region as a result of pulse excitation in an external electric field. The excitation was provided by unpolarized light pulses of $2 \cdot 10^{-4}$ sec duration and $\lambda_{exc} = 560$ nm wavelength. At 300°K the luminescence decay constant was 3.2 msec but at 77°K it rose to 7.6 msec.

This rise in τ was due to an increase in the reabsorption of the luminescence. In fields $\mathscr{E} > 50$ kV/cm the value of τ decreased continuously with increasing field and it amounted to 15% in a field of 225 kV/cm (Table 7). It was interesting to note that in weak fields $\mathscr{E} < 50$ kV/cm the value of τ increased somewhat. This increase was undoubtedly due to an increase in the reabsorption in the R-line region and it was evidence of an absolute increase in the sorption coefficient at the line center due to the application of weak fields. If in weak fields the absorption integral remained constant $(f_{\mathscr{E}}/f_0 = 1)$, i.e., if the lines were broadened and this was accompanied by an increase in the absorption at the line centers, the reabsorption and, consequently, τ should decrease.

The decay curves were used in estimating the quantum efficiency in the presence of the electric field. For a constant number of absorbed photons the integral of the decay curve was proportional to the quantum efficiency:

$$\int_0^\infty I_0 \exp\left(-\frac{t}{\tau}\right) dt = I_0 \tau \propto \eta, \tag{12}$$

where I_0 was the initial luminescence intensity. When the field intensity was increased, the efficiency $\eta_{\mathscr{E}} \propto I_0 \tau_{\mathscr{E}}$ increased continuously and reached $\eta_{\mathscr{E}} = 1.12 \eta_0$ in a field of 225 kV/cm. The rise of $\eta_{\mathscr{E}}$ with $\mathscr{E}$ was largely due to a redistribution of the intensity in the luminescence spectrum of ruby because of a reduction in the reabsorption and could not be attributed entirely to the increase in the quantum efficiency of ruby. However, it should be noted that the greatest changes in $\tau_{\mathscr{E}}$ and $\eta_{\mathscr{E}}$ were observed in fields $\mathscr{E} > 150$ kV/cm, when the change in the reabsorption should be slight.

An increase in the intensity of the $^4A_2 \rightarrow {}^2E$ transition and reduction in τ were also observed for a sample with a lower value of $C(Cr^{3+})$ in a field $\mathscr{E} < 50$ kV/cm.

TABLE 7. Influence of External Electric Field $\mathscr{E} \parallel C_3$ on Duration of Decay of Ruby Luminescence (77°K)

$\mathscr{E}$, kv/cm	$\tau_{\mathscr{E}}$, msec	$\tau_{\mathscr{E}}/\tau_0$	$\mathscr{E}$, kV/cm	$\tau_{\mathscr{E}}$, msec	$\tau_{\mathscr{E}}/\tau_0$	$\mathscr{E}$, kV/cm	$\tau_{\mathscr{E}}$, msec	$\tau_{\mathscr{E}}/\tau_0$
0	7.6±0.2	1	120	7.18	0.945	200	6.70	0,882
50	7,93	1.04	160	7.16	0.942	208	6.58	0,865
70	7.23	0.952	170	7,11	0.935	225	6.48	0,85
100	7.21	0.949	188	6.84	0.90			

§ 3. Discussion of Results

As expected, an anisotropic perturbation by the $\mathscr{E} \parallel C_3$ field increased the probability of the $^4A_2 \rightleftarrows {}^2E$ transitions. The intensities of the $^4A_2 \rightarrow {}^4T_2$, a^4T_1 transitions were not affected by the external field and the intensity of the $^4A_2 \rightarrow a^2T_2\,(\underline{\overline{E}}_a)$ transition even decreased. An external field $\mathscr{E} \parallel C_3$ did not reduce the crystal symmetry but altered only the trigonal component V_{trig}, whereas local fields of color centers were capable of reducing the symmetry of the impurity centers to the triclinic form. The inequivalence of the action of the local fields of color centers and of an external axial field was the cause of the different effects of the perturbation of the crystal field on the intensities of the forbidden transitions.

The different influence of the perturbing $\mathscr{E} \parallel C_3$ field on the intensities of the absorption in the U band and R lines indicated that there was no linear relationship between the intensities of the corresponding transitions, though such a relationship was postulated in [2]. The insensitivity of the U band to the field indicated that V_{trig} was not the interaction that governed the probability of the $^4A_2 \rightarrow {}^4T_2$ transition. The large discrepancy between the experimental results and the theoretical estimates of the probabilities of transitions in the Cr^{3+} ions and the optical anisotropy of these transitions also suggested that the trigonal component of the crystal field was not the main factor governing the intensity of the forbidden transitions. Among all the intercombination transitions in ruby the resonance transition had the lowest intensity and, therefore, the change in V_{trig} as a result of perturbation with the field $\mathscr{E} \parallel C_3$ had a stronger influence on the probability of the resonance transition.

A strong increase in the intensity of the B lines in colored ruby and the converse effect due to the anisotropic perturbation by the $\mathscr{E} \parallel C_3$ field indicated that the intensity of this transition was govered by perturbations of lower symmetry than C_{3v} and we could expect an increase in the intensity of the $^4A_2 \rightarrow {}^2T_2$ transitions as a result of application of a field with the $\mathscr{E} \perp C_3$ orientation.

Thus, the different influence of the electric field $\mathscr{E} \parallel C_3$ on the probabilities of the transitions in question raised doubts about the dominant role of V_{trig} in the mechanism responsible for the lifting of the parity forbiddenness. A similar inclusion could be also drawn about the role of the spin—orbit interaction V_{so} in the lifting of the intercombination forbiddenness. If the intracenter spin—orbit interaction were the only cause of the lifting of the forbiddenness of the intercombination transitions, one would expect the same changes in the probabilities of the $^4A_2 \rightarrow {}^2E$ and $^4A_2 \rightarrow {}^4T_2$ transitions. A considerable increase in the intensity of the resonance transition in the field $\mathscr{E} \parallel C_3$ and the insensitivity of the U band to the same field indicated that there were other factors which contributed to the lifting of the intercombination forbiddenness and these other factors could be the main ones. For example, the forbiddenness could be lifted mainly because of the interactions with ligands, i.e., because of the covalent interaction. This interaction was manifested clearly by a strong increase in the intensity of the intercombination transitions to upper states in the doublet system (K band) with which the ^2E state interacted more strongly than with the quartet states. As shown in [3], the initial splittings of the $^4A_2$, ^2E, and $^2T_2$ terms were very sensitive to the covalent interaction parameters, whereas attempts to explain the initial splittings allowing only for V_{trig} and V_{so} were unsuccessful. This also supported the predominance of the covalent interaction.

A preliminary conclusion that $\eta\,(R)$ increased at 77°K, deduced from an increase in the probability of the resonance transition and radiative efficiency of the R lines in the external field $\mathscr{E} \parallel C_3$, would require a confirmation because the result was obtained for a sample with a low quantum efficiency. This confirmation could be obtained by investigating the influence of the external field on $\eta\,(R)$, $\eta\,(U)$, and $\eta\,(Y)$ (and not on the radiative efficiency of the R lines) in ruby crystals of higher quality.

CONCLUSIONS

The main results of this investigation can be summarized as follows.

1. An investigation was made of the spectral-luminescence characteristics of ruby. An activator absorption band near the fundamental edge (K band) was discovered and interpreted. A low-temperature fall of the luminescence efficiency was observed under resonance excitation conditions and the mechanism of the intracenter thermal quenching of the luminescence was determined more accurately.

2. Recombination luminescence of ruby excited by high-power light pulses was observed and investigated. It was found that the phosphorescence involving the ^2E state as an intermediate level was excited in two stages. High-power optical excitation could reduce the concentration of the chromium activator and could produce the Cr^{4+} centers. The results obtained were used to construct an energy-band model of ruby postulating the presence of a considerable energy shift between the optical and minimal widths of the forbidden band.

3. A study was made of the nature of the color centers formed in ruby as a result of high-power optical or γ-ray irradiation. In both cases the dominant color centers were the Cr^{2+} and Cr^{4+} ions. The role of the Ti^{3+} impurity in both types of coloration was considered. The absorption bands associated with the color centers were identified.

4. An increase in the intensity of the intercombination transitions in Cr^{3+} ($^4A_2 \rightarrow \, ^2E, \, ^2T_2$) in the presence of color centers was investigated and interpreted. A relationship between the formation of color centers in a ruby laser element and the laser characteristics was established.

5. Investigations were made of the intensities of the intracenter transitions $^4A_2 \rightarrow \, ^4T_2$, a^4T_1, ^2E, $^2T_2$, in an external electric field applied along the trigonal axis. A considerable increase in the probability of the $^4A_2 \rightleftharpoons \, ^2E$ transitions and of the radiative efficiency of the R lines was observed when a crystal was perturbed anisotropically by an external field $\mathscr{E} \parallel C_3$. The intracenter spin−orbit interaction was not the dominant factor governing the resonance transition intensity. It was postulated that the covalent interaction played the dominant role in the lifting of the intercombination and parity forbiddenness of the $^4A_2 \rightleftharpoons \, ^2E$ transitions.

The results of the spectral-luminescence investigations taken as a whole helped to provide a better understanding of the processes occurring in ruby under high-power optical excitation conditions, and the causes of the instability of the more important ruby laser characteristics during prolonged operation were established. The basic similarity between the quantum-mechanical systems used in solid-state lasers suggested that some of the processes and phenomena occurring in ruby (optical coloration mechanism of changes in the transition probabilities in the presence of color centers) should occur also in other solid-state lasers.

The author has a pleasant duty to express his sincere thanks to Z. L. Morgenshtern for suggesting the problem and directing this work, to M. D. Galanin, V. V. Antonov-Romanovskii, and A. M. Leontovich for valuable discussions and advice, to G. E. Arkhangel'skii for helpful cooperation, and B. M. Anfinogenov for his constant help.

LITERATURE CITED

1. D. S. McClure, J. Chem. Phys., 36:2757 (1962); 38:2289 (1963).
2. S. Sugano and Y. Tanabe, J. Phys. Soc. Jap., 13:880 (1958).
3. S. Sugano and M. Peter, Phys. Rev., 122:381 (1961).
4. N. K. Bel'skii and D. A. Mukhamedova, Dokl. Akad. Nauk SSSR, 158:317 (1964); 162:527 (1965).
5. J. Margerie, C. R. Acad. Sci. (Paris), 255:1598 (1962).
6. S. Sugano and I. Tsujikawa, J. Phys. Soc. Jap., 13:899 (1958).

7. G. K. Klauminzer, P. L. Scott, and H. W. Moos, Phys. Rev., 142:248 (1966).

8. B. Z. Malkin, Zh. Eksp. Teor. Fiz., 42:1410 (1962); Fiz. Tverd. Tela, 4:2214 (1962).

9. S. A. Pollack, IEEE J. Quantum Electron., 4:703 (1968).

10. E. E. Bukke and Z. L. Morgenshtern, Opt. Spektrosk., 14:687 (1963); Acta Phys. Pol., 26:593 (1964).

11. G. V. Schultz, Z. Phys., 167:446 (1962).

12. M. D. Galanin, V. N. Smorchkov, and Z. A. Chizhikova, Opt. Spektrosk., 19:296 (1965).

13. B. S. Tsukerblat and Yu. E. Perlin, Opt. Spektrosk., 21:13 (1966).

14. Z. L. Morgenshtern and V. B. Neustruev, Opt. Spektrosk., 20:837 (1966).

15. E. I. Al'shits, Z. L. Morgenshtern, and V. B. Neustruev, Opt. Spektrosk., 31:932 (1971).

16. K. Watanabe and E. C. Y. Inn, J. Opt. Soc. Am., 43:32 (1953).

17. S. V. Grum-Grzhimailo, L. B. Pasternak, D. T. Sviridov, L. G. Chentsova, and M. A. Chernysheva, in: Spectroscopy of Crystals [in Russian], Nauka, Moscow (1966), p. 168.

18. E. Loh, J. Chem. Phys., 44:1940 (1966).

19. E. T. Arakawa and M. W. Williams, J. Phys. Chem. Solids, 29:735 (1968).

20. D. T. Sviridov, Opt. Spektrosk., 20:488 (1966).

21. D. T. Sviridov, Dokl. Akad. Nauk SSSR, 157:334 (1964).

22. T. Kushida, J. Phys. Soc. Jap., 21:1331 (1966).

23. Z. L. Morgenshtern and V. B. Neustruev, Opt. Spektrosk., 32:953 (1972); in: Ruby and Sapphire [in Russian], Nauka, Moscow (1974), p. 130.

24. Z. L. Morgenshtern and V. B. Neustruev, Izv. Akad. Nauk SSSR Ser. Fiz., 37:641 (1973).

25. T. H. Maiman, Phys. Rev. Lett., 4:564 (1960).

26. Z. L. Morgenshtern and V. B. Neustruev, ZhETF Pis'ma Red., 2:507 (1965); Phys. Status Solidi, 14:303 (1966).

27. V. V. Antonov-Romanovskii, Dokl. Akad. Nauk SSSR, 2(11):93 (1936).

28. T. S. Moss, Optical Properties of Semiconductors, Butterworths, London (1959).

29. P. J. Harrop and R. H. Creamer, Br. J. Appl. Phys., 14:335 (1963).

30. O. T. Ozkan and A. J. Moulson, J. Phys. D, 3:983 (1970).

31. T. Maruyama and Y. Matsuda, J. Phys. Soc. Jap., 19:1096 (1964).

32. D. R. Mason and J. S. Thorp, Proc. Phys. Soc. Lond. 87:49 (1966).

33. C. A. Bates and J. M. Dixon, J. Phys. C, 2:2225 (1969).

34. G. V. Schultz, Phys. Lett., 9:301 (1964).

35. S. F. Sharlai, Postgraduate Papers [in Russian], Leningrad Institute of Precision Mechanics and Optics (1966), p. 3.

36. C. M. Stickely, H. Miller, E. E. Hoell, C. C. Gallagher, and R. A. Bradbury, J. Appl. Phys., 40:1792 (1969).

37. W. Flowers and J. Jenney, Proc. IEEE, 51:858 (1963).

38. G. E. Arkhangelskii, Z. L. Morgenshtern, and V. B. Neustruev, Phys. Status Solid, 22: 289 (1967).

39. G. E. Arkhangelskii, Z. L. Morgenshtern, and V. B. Neustruev, Phys. Status Solidi, 29: 831 (1968).

40. G. E. Arkhangelskii, Z. L. Morgenshtern, and V. B. Neustruev, in: Spectroscopy of Crystals [in Russian], Nauka, Moscow (1970), p. 273.

41. V. V. Antonov-Romanovskii, Kinetics of Photoluminescence of Crystal Phosphors [in Russian], Nauka, Moscow (1966), § § 36 and 38.

42. L. S. Kornienko and A. M. Prokhorov, Zh. Eksp. Teor. Fiz., 38:1651 (1960).

43. D. S. McClure, Solid State Phys., 9:399 (1959).

44. D. T. Sviridov, Krist. Tech., 3:661 (1968).

45. Z. L. Morgenshtern and V. B. Neustruev, Izv. Akad. Nauk SSSR, Ser. Fiz., 32:6 (1968).

46. W. R. Davis, A. S. Menius, Jr., M. K. Moss, and C. R. Philbrick, J. Appl. Phys., 36:670 (1965).

47. V. R. Johnson and R. W. Grow, IEEE J. Quantum Electron., QE-3:1 (1967).
48. G. E. Arkhangel'skii, Z. L. Morgenshtern, and V. B. Neustruev, Izv. Akad. Nauk SSSR, Ser. Fiz., 33:807 (1969); Phys. Status Solidi, 36:451 (1969).
49. A. I. Ritus and A. A. Manenkov, Fiz. Tverd. Tela, 5:3590 (1963).
50. V. A. Gevorkyan, K. A. Madatyan, É. A. Kochinyan, and V. Kh. Sarkisov, in: Spectroscopy of Crystals [in Russian], Nauka, Moscow (1970), p. 280.
51. V. B. Neustruev, Zh. Prikl. Spektrosk., 12:948 (1970).
52. Z. L. Morgenshtern and V. B. Neustruev, Spectroscopy of Crystals [in Russian], Nauka, Leningrad (1973), p. 291.
53. W. Kaiser, S. Sugano, and D. L. Wood, Phys. Rev. Lett., 6:605 (1961).
54. M. G. Cohen and N. Bloembergen, Phys. Rev., 135:A950 (1964).
55. A. A. Kaplyanskii and V. N. Medvedev, Fiz. Tverd. Tela, 9:2704 (1967).
56. G. Burns and M. I. Nathan, J. Appl. Phys., 34:703 (1963).

RELATIONSHIP BETWEEN OPTICAL AND THERMAL DEPTHS OF ELECTRON TRAPS

Wu Kuang and M. V. Fok

A method is suggested for the determination of the optical depth of electron traps from the position of the exponential part of the decay curve of the flash sensitivity of a phosphor, considered as a function of the energy of infrared radiation used to illuminate the phosphor. This method is used to demonstrate that the difference between the optical and thermal depths of traps is independent of their depth and it is either 0.6 or 0.3 eV. Both these values may be manifested by the same trap depending on whether the liberation of electrons from the trap is adiabatic or nonadiabatic. It is also shown that the difference between the optical and thermal depths is related to the presence of ions in the crystal lattice because this difference vanishes in the case of covalent crystals.

Introduction
___

The depth of an electron or hole trap is a measure of the energy necessary for its ionization, i.e., for the liberation of the charge localized in it. However, the thermal energy needed for the ionization of a trap is frequently significantly different from the energy required for the optical ionization of the same trap. This difference arises because the thermal and optical liberation of charges from traps differs not only in respect of the mechanism, but also in respect of the possible final states of the system. Therefore, the very concept of a trap can be introduced only for a definite mechanism of liberation of charges localized in it.

In particular, the thermal depth of a trap is usually identified with the activation energy which governs the temperature dependence of the probability of charge liberation. This energy is evolved when a free charge, which is in thermal equilibrium with the crystal lattice, becomes localized in a given trap. The activation energy is equal to the difference between the free energies of the system with free and localized electrons.

The optical depth of a trap is equal to the minimum photon energy sufficient to liberate the localized charge from the trap without participation of the thermal vibrations.

The aim of the present paper is to attempt to determine the cause of the difference between the optical and thermal depths of a trap and to find the relationship between them.

It is suggested in [1] that the difference between the optical and thermal depths of a trap is due to the polarization of the lattice in the field of a charge being localized (or liberated).*

___
* In principle, we can distinguish two cases: In one case a trap is neutral before the localization of a charge and in the other it becomes neutral after localization. In the first case the lattice is polarized after the capture of a charge and in the second case after its liberation.

39

This polarization appears when there is a sufficient number of ions in the crystal lattice, for example, if the chemical binding in a given compound is partially ionic or when the compound contains a large number of mutually compensated donors and acceptors. These are precisely the conditions for the appearance of polarons. The displacement of ions due to the appearance of a polaron is essentially the same as the displacement due to the polarization of the lattice by a charged trap. Therefore, the energies evolved in these two cases should be approximately the same. We shall assume that the Franck—Condon principle is obeyed in the optical liberation of charges from traps: According to this principle the coordinates and velocities of atoms or ions do not change in a time needed for an electron transition. This means that after a transition such atoms or ions are in a nonequilibrium state and the polarization mentioned above occurs after the transition. When an electron is liberated, there is a change in the charge at two points in the crystal lattice: near the trap from which the electron reaches after the transition and where it interacts for the first time with a phonon. This point may be separated by several tens and even a hundred lattice constants from the trap. Consequently, the lattice polarization and energy evolution should also occur at two points, and the total energy evolved should be equal to twice the polaron binding energy.

It follows from these considerations that the difference between the optical and thermal depths of a given trap should be independent of its depth and should be the same as the difference between the "optical" and "thermal" values of the forbidden-band width of a given crystal. This point is proved in [1] for deep traps in zinc sulfide. However, it is not clear whether the same considerations apply also to shallow traps because in this case a localized charge may describe an orbit of fairly large radius, and, consequently, it may polarize only weakly the crystal lattice. Therefore, we shall be concerned primarily with shallow traps in the same crystal phosphor ZnS:Cu:Co:Cl, which was used in [1]. However, we shall adopt a much improved investigation method.

The method for the determination of optical and thermal trap depths should make it possible to establish reliably that the values obtained apply to the same trap.

There are many experimental methods for the determination of thermal depths of traps. One of the most widely used is the thermoluminescence method. However, it is inapplicable to the present case because the depth obtained cannot be compared with the optical value. The latter can be found by illuminating a phosphor with radiation of sufficiently long wavelengths and finding the minimum photon energy which is just sufficient to liberate the trapped charge. The actual liberation of a charge from a trap is manifested by luminescence stimulation (flash) or quenching under the influence of long-wavelength radiation, provided this influence is sufficiently strong. Therefore, the optical trap depth is frequently identified with the maximum of sensitivity to long-wavelength radiation. However, it follows from the foregoing comments that this is incorrect and we have to find experimentally not the maximum value but the long-wavelength sensitivity edge.

If the optical depth of a trap is known, the thermal depth can be determined by investigating the temperature dependence of the stimulating (flash) or quenching action of long-wavelength radiation. The basis of the proposed method can be found in [2, 3].

§ 1.   Theory of the Method

Our experimental study is confined to electron traps (and the corresponding stimulating influence of infrared radiation). Therefore, we shall present the theory of this case.

Let us consider the energy-band scheme of a crystal phosphor with q electron and r hole traps, which can act also as recombination centers and some as luminescence centers (Fig. 1). The system of rate equations applicable to this case is considered, for example, in

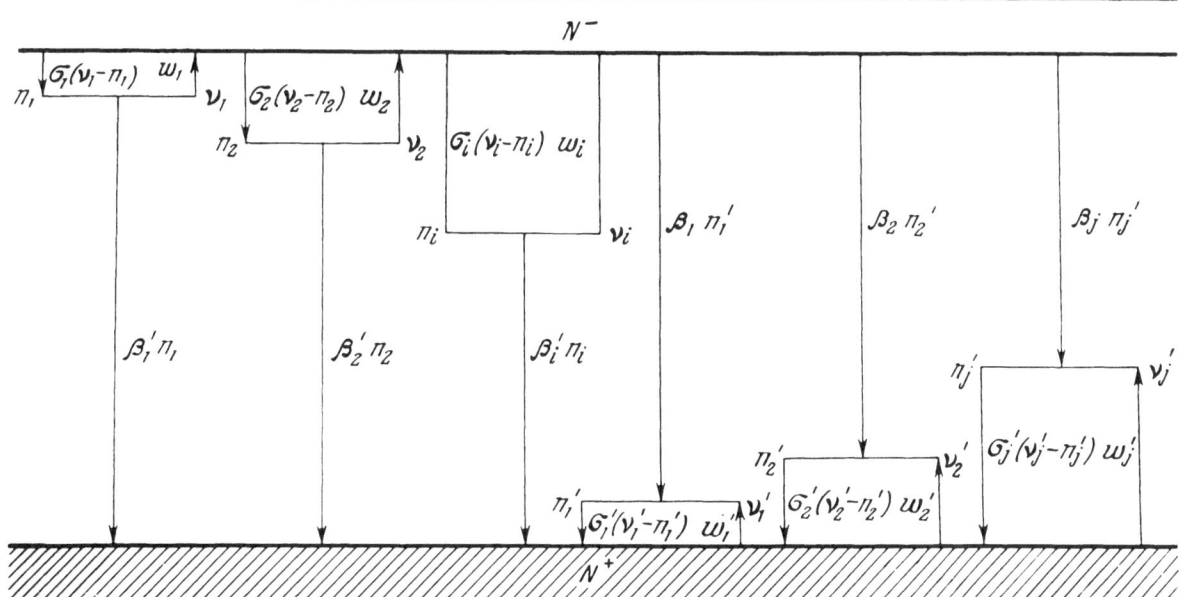

Fig. 1. Energy-band scheme of a crystal phosphor. The arrows represent electron transitions. The corresponding probabilities are given alongside the arrows. The horizontal lines represent trap levels. The trap concentrations are given on the right of each horizontal line and the densities of localized electrons (in the case of electron traps) or holes (in the case of hole traps) are given on the left of each line.

the book of Antonov-Romanovskii [4]. We shall express this system in a somewhat more general form so as to clarify later the restrictions which are made:

$$
\begin{aligned}
\frac{dn_i}{dt} &= \sigma_i u^- (\nu_i - n_i) N^- - w_i n_i - \beta_i' N^+ n_i \quad (i = 1, 2, \ldots, q), \\
\frac{dn_j'}{dt} &= \sigma_j' u^+ (\nu_j' - n_j') N^+ - w_j' n_j' - \beta_j N^- n_j' \quad (j = 1, 2, \ldots, r), \\
\frac{dN^-}{dt} &= \sum_{i=1}^{q} w_i n_i - u^- \sum_{i=1}^{q} \sigma_i (\nu_i - n_i) N^- - N^- \sum_{j=1}^{r} \beta_j n_j', \\
N^- &+ \sum_{i=1}^{q} n_i = N^+ + \sum_{j=1}^{r} n_j' + N_d - N_a.
\end{aligned}
\tag{1}
$$

Here, n_i and n_j' are the densities of electrons and holes localized at i-th electron traps and j-th hole traps; σ_i and σ_j' are the carrier-capture cross sections of the traps; ν_i and ν_j' are the concentrations of the traps; w_i and w_j' are the probabilities of liberation of charges from the traps; u^- and u^+ are the thermal velocities of free electrons and holes; N^- and N^+ are the densities of free electrons and holes; β_i is the recombination coefficient of a free hole with an electron localized at a trap i; β_j is the recombination coefficient of a free electron with a hole localized at a trap j. The dimensions of the coefficient β are the same as the dimensions of the product σu. The symbols N_d and N_a denote, as usual, the donor and acceptor concentrations.

The system (1) describes the afterglow. Therefore, it does not include the intensity of the luminescence-exciting light. This is important because the exciting light can itself liberate charges from traps and thus influence the values of w.

If a certain time is allowed after the end of excitation, so that the fast processes are completed, the density of free electrons varies in the same way as the density of localized electrons. The lifetime of free electrons in the conduction band is usually not more than 10^{-9} sec, and the thermalization time of "hot" electrons which may be generated by the light is even shorter. Therefore, we may assume that the quasiequilibrium density of free electrons is established in a time shorter than 10^{-7} sec. Mathematically, this means that in the equation for N^- in the system (1) we can drop the derivative with respect to time and assume that this equation is algebraic.* We can then easily find the value of N^-:

$$N^- = \frac{\sum_{i=1}^{q} w_i n_i}{u - \sum_{i=1}^{q} \mathfrak{z}_i (\nu_i - n_i) + \sum_{j=1}^{r} \beta_j n'_j}.$$ (2)

The brightness J_j of the luminescence due to the recombination of free electrons with holes localized at traps of type j is proportional to the rate of recombination at these traps:

$$J_j \propto \beta_j n'_j N^-.$$ (3)

When a phosphor is then illuminated with long-wavelength radiation of photon energy $h\nu$, the probability of liberation of electrons from all traps increases:

$$w_{i\ \text{IR}}(h\nu) = w_i + p_i(h\nu) I_{h\nu},$$ (4)

where $I_{h\nu}$ is proportional to the intensity of infrared radiation expressed as the number of photons and $p_i(h\nu)$ is the probability of liberation of an electron from a level i under the influence of one photon of energy $h\nu$. An increase in w_i results, in accordance with Eq. (2), in an increase in N^- and, therefore, in the luminescence brightness. Soon after the beginning of infrared illumination the densities of localized charges cease to change significantly. Therefore, the brightness of all the luminescence bands rises by the same factor. It follows from Eqs. (2) and (3) that the relative amplitude of a "flash" stimulated in this way is

$$R \equiv \frac{J_{j,\text{IR}} - J_j}{J_j} = \frac{\sum_{i=1}^{q} p_i(h\nu) n_i}{\sum_{i=1}^{q} w_i n_i} I.$$ (5)

It is clear from the above formula that R is independent of the degree of occupancy of the traps (of the relationship between ν_i and n_i), and of any external quenching (this quantity does not include any recombination terms such as $\beta'_i n_i$ or $\beta_j n'_j$). Therefore, if the conditions are such that the optical liberation of electrons occurs mainly from levels of one kind, the function $R(h\nu)/I_{h\nu}$ represents (in relative units) the dependence $p(h\nu)$ for these levels. We shall now establish under what conditions this can happen.

When the energy of the incident quanta is reduced and $h\nu$ becomes just smaller than the optical trap depth $E_{i\ \text{opt}}$, the value of $p_i(h\nu)$ begins to decrease exponentially

$$p_i(h\nu) \propto \exp\left(\frac{E_{\text{opt}} - h\nu}{kT}\right).$$ (6)

This formula has a simple physical meaning. The difference $E_{\text{opt}} - h\nu$ is simply the thermal energy which must be added to the energy of an infrared photon for the liberation of an electron

---

* The derivatives cannot be dropped from the other equations of the system (1).

from a trap. The probability that an electron receives this additional energy is proportional to the Boltzmann factor. Under the conditions discussed here all transitions occur to the lowest level in the conduction band irrespective of the value of $h\nu$. Therefore, the other factors in the expression for the probability of an electron transition, representing the wave functions of the initial and final states, can be regarded as independent of $h\nu$. [If $h\nu > E_{opt}$, the dependence $p(h\nu)$ is governed by these factors.]

It is clear from Eq. (6) that when $h\nu$ is small, the quantity $p(h\nu)$ may decrease by a large factor compared with its maximum value. Therefore, if we select $h\nu$ so that it is smaller than E_{opt} for all the traps, except the shallowest, we place the latter in a preferential position. If, moreover, we begin infrared illumination at the beginning of the afterglow when the number of electrons in shallow traps is still practically unchanged, we may ensure that

$$p_1(h\nu) n_1 \gg \sum_{i=2}^{q} p_i(h\nu) n_i \qquad (7)$$

(we shall label traps in increasing order of their depth). The condition (7) is easiest to satisfy by exciting a phosphor with short light pulses so that the occupancy of the traps is governed by their capture cross sections and not by their depths.

If we increase $h\nu \geq E_{2\,opt}$, electrons are liberated preferentially from the shallowest and second-shallowest traps. Starting infrared illumination during later stages of the afterglow we can find the moment when the shallowest traps are already empty and the second-shallowest have still significant numbers of electrons. Then, Eq. (5) is dominated by the second term. We can proceed similarly in the selection of the third and higher terms. However, we may find that the required conditions are obeyed only in a narrow range of the photon energy $h\nu$, which is insufficient to find reliably a region with a slope equal to unity. Then, we can eliminate approximately the influence of shallower traps by substracting from all values of R/I the value corresponding to the horizontal part of a step at longer wavelengths. This correction is fully justified if $p(h\nu)$ is constant for $h\nu > E_{opt}$. In fact, this value varies slowly (compared with the region where $h\nu > E_{opt}$) and has a flat maximum. If the region with a slope equal to unity does indeed fall within the range of $h\nu$ under consideration, the value of R/I varies so rapidly that the correction in question may influence only the part of the curve at the longest wavelengths. Nevertheless, this is still useful because the required rectilinear region becomes more prominent.

§2. Phototransitions and Associated

Optical Trap Depths

We have mentioned earlier that the optical depth of electron traps is governed by the minimum value of the photon energy capable of liberating an electron from a trap without participation of the thermal energy of the atomic vibrations.

We shall now consider how we can determine E_{opt} for each trap. We shall assume that we have already found experimentally the conditions under which the numerator of Eq. (5) is dominated by the term corresponding to this trap and, consequently, the spectral dependence $R(h\nu)/I(h\nu)$ gives the dependence $p(h\nu)$. We then determine the dependence of $R(h\nu)/I(h\nu)$ on $h\nu$, and we plot $\ln(R/I)$ as a function of $h\nu/kT$. In the region where Eq. (6) is valid we obtain a straight line with a slope equal to unity. If $h\nu \geq E_{opt}$, this dependence should be less steep. Therefore, the optical depth of a trap can be regarded as equal to the energy of the photons corresponding to the short-wavelength end of this linear region.

Having determined the optical depth, we can then measure the thermal depth of the same traps by a method based on the temperature dependence of the flash (stimulation effect) using photons corresponding exactly to the optical depth of the traps to liberate electrons. Clearly, in this case there is no need for an additional experimental proof that the optical and thermal depths obtained by these methods apply to the same type of trap.

We shall now consider the method which can be used to determine the thermal depth of traps. In this case we require somewhat more stringent experimental conditions because restrictions in Eq. (6) apply not only to the numerator, but also to the denominator. If the term corresponding to traps of a given kind predominates not only in the numerator of this formula, but also in its denominator, we find that it changes to

$$\frac{R}{I_{h\nu}} = \frac{p_i(h\nu)}{w_i} = \frac{p_i(h\nu)}{w_{i0}} \exp\left(\frac{E_{i,\text{th}}}{kT}\right),$$ (8)

since the probability of thermal liberation of electrons is an exponential function of temperature. This allows us to determine E_{th}. If we find the dependence $R(T)$ for constant values of I and $h\nu$ (and then plot it using the coordinates $\ln R$ and I/T), we obtain a straight line whose slope is proportional to E_{th}. However, this simple case is only rarely encountered in practice. This is due to the fact that liberation of carriers from shallow traps ends when they reach equilibrium with the next deeper traps, i.e., when

$$\frac{w_1 n_1}{\sigma_1 v_1} = \frac{w_2 n_2}{\sigma_2 v_2}.$$ (9)

This relationship is obtained from the system (1) if we assume that

$$\frac{dn_1}{dt} \ll w_1 n_1, \qquad \beta_1' N^+ \ll w_1,$$

$$\frac{dn_2}{dt} \ll w_2 n_2, \qquad \beta_2' N^+ \ll w_2,$$ (10)

i.e., if we assume that the trap occupancy is not governed by recombination, but by exchange of electrons with the conduction band. Moreover, we have ignored here n_1 and n_2 compared with ν_1 and ν_2, since we are speaking of fairly late stages of the afterglow. If $\sigma_1 \nu_1$ and $\sigma_2 \nu_2$ are of the same order of magnitude, then $w_1 n_1$ becomes of the same order of magnitude as $w_2 n_2$. Nevertheless, this does not interfere with the determination of the thermal depth of traps but, on the contrary, it extends the capabilities of the method. In fact, we shall assume that during the afterglow stage under consideration, Eq. (9) is already satisfied and at the same time we have

$$\sum_{i=3}^{r} w_i n_i \ll w_1 n_1 + w_2 n_2.$$ (11)

If the depth of the third trap exceeds the depth of the second by 10 kT or more, the above inequality is usually satisfied during a certain state of the afterglow. In this case only the first two terms remain in the denominator of Eq. (5) and it follows from Eq. (9) that Eq. (5) then becomes

$$w_1 n_1 + w_2 n_2 = \left(1 + \frac{\sigma_2 v_2}{\sigma_1 v_1}\right) w_1 n_1 = \left(1 + \frac{\sigma_1 v_1}{\sigma_2 v_2}\right) w_2 n_2.$$ (12)

Hence, it is clear that, depending on the photon energy, we can measure the thermal depth of the first or second level. In fact, if $h\nu$ is selected so that the condition (7) is satisfied, we can rewrite Eq. (5) in the form

$$R = \frac{p_1(h\nu)}{\left(1 + \frac{\sigma_2\nu_2}{\sigma_1\nu_1}\right)w_{10}} \exp\left(-\frac{E_{1,\text{th}}}{kT}\right) I. \tag{13}$$

However, if the experimental conditions are selected so that infrared radiation liberates electrons from the second-shallowest traps, the same formula can be written differently:

$$R = \frac{p_2(h\nu)}{\left(1 + \frac{\sigma_1\nu_1}{\sigma_2\nu_2}\right)w_{20}} \exp\left(-\frac{E_{2,\text{th}}}{kT}\right) I. \tag{14}$$

Thus, we can determine the depth of the shallowest, second-shallowest, and other traps. We have to find suitable experimental conditions, i.e., the range of temperatures and the afterglow stage for which the conditions such as those in Eq. (10) are satisfied for k-th traps. We can readily see that in this case the expression for R is similar to that of Eq. (14) except that $[1 + (\sigma_1\nu_1/\sigma_2\nu_2)]$ should be replaced with $\sum_{i=1}^{\kappa} \frac{\sigma_i\nu_i}{\sigma_\kappa\nu_\kappa}$.

Formulas such as Eq. (13) yield also the experimental criterion that the conditions for finding E_{th} are correctly selected. If a graph does not have a rectilinear region, it follows that the thermal equilibrium between traps from which significant liberation of electrons takes place has not yet been reached. In this case it is useful to study a later stage of the afterglow or use higher temperatures. In measuring the depth of shallow traps it is preferable to go over to earlier stages of the afterglow or to lower temperatures.

§ 3. Apparatus and Measurement Procedure

We used a special brass cryostat, which — together with auxiliary apparatus — enabled us to establish and maintain for a long time any temperature between 80 and 373°K and to avoid precipitation of moisture.

The cryostat consisted of three main parts: a pillar, main body, and reservoir for liquid nitrogen (Fig. 2). The upper part of the pillar was a double-walled brass cylinder. The lower part of its inner wall was attached to a copper can which had a holder with three recesses for samples. A Nichrome heater and a copper resistance thermometer were wound on the can. A thermocouple was soldered into one of the recesses. The copper can was screwed onto the inner wall of the brass cylinder. Connecting wires from the heater, thermocouple, and resistance thermometer passed through a special vacuum seal in the outer wall of the brass cylinder.

The lower part of the pillar A fitted inside the main body B. The outer wall of the pillar was clamped tightly against the body via a rubber spacer. The side wall of the main body had three windows covered with quartz, glass, and fluorite windows for the transmission of exciting ultraviolet light, visible luminescence, and deexciting infrared radiation. An additional glass window was fitted in the cover of the main body, and it was used to view a sample visually during the assembly. This window was covered by an opaque screen during measurements. An opening with a connection to a pump was fitted in the bottom of the main body. The pressure inside the cryostat was maintained at about 10^{-2} mm Hg. A liquid nitrogen reservoir C fitted inside the pillar A. The upper part of the reservoir was made of Textolite and the lower one, of copper.

A phosphor powder layer about $2 \cdot 10^{-3}$ cm thick was precipitated on aluminum substrates. These substrates (or, in some cases single crystals of the phosphor) were held by springs in the recesses or they were soldered with Wood's alloy. Samples were excited with ultraviolet light of $\lambda = 365$ nm or 312 nm wavelengths separated from the spectrum of a PRK-4 lamp with an ultraviolet filter and an aqueous solution of $CuSO_4$ or $NiSO_4$ in a quartz cell.

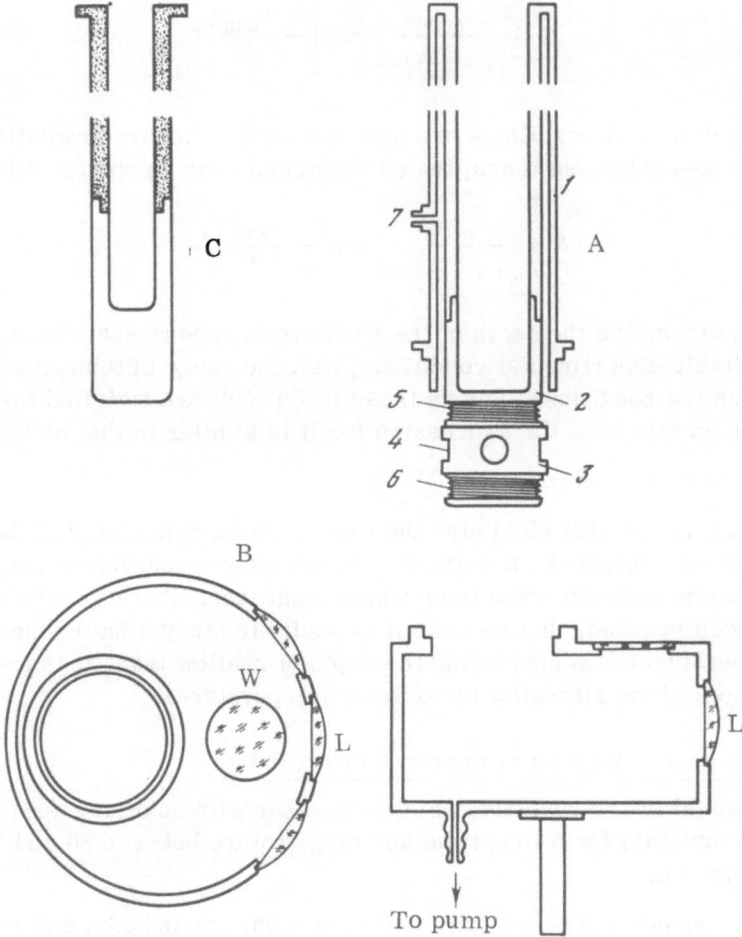

Fig. 2. Cryostat. Here A is the pillar, B is the main body (the view from above is shown on the left and a section on the right), and C is the liquid nitrogen reservoir; 1) double-walled brass cylinder; 2) copper can; 3) sample holder; 4) recess for sample; 5) heater; 6) resistance thermometer; 7) sealed aperture for leads; L are lenses and W is a window.

The intensity of the ultraviolet exciting light could be varied by neutral filters and grids. Infrared radiation was provided by an incandescent lamp with a sapphire window, which was supplied through a ferroresonance stabilizer, laboratory autotransformer, and current transformer. Monochromatic infrared radiation was selected by a combination of two interference filters and a germanium or silicon absorption filter. Some of the characteristics of the interference filters were as follows:

Maximum transmission, %	40-70
Half-width, μ	0.05-0.11
Width of pass band, μ	0.17-0.70
Transmission outside pass band, %	≤ 0.5- ≤ 1.5

A combination of two interference filters with similar transmission maxima reduced the maximum transmission by a factor of about 2 and the half-width varied depending on the relative

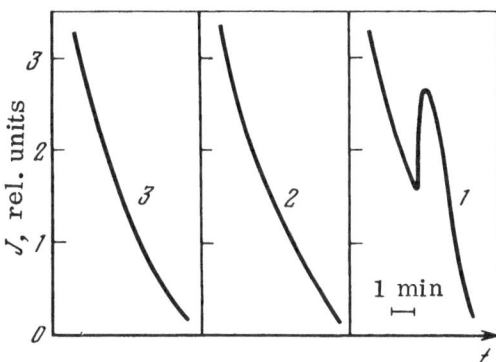

Fig. 3. Influence of infrared radiation on the afterglow of ZnS:Cu:Co:Cl: 1) flash produced by radiation passed through the 2.08 + 2.13 μ filter combination; 2) same as 1 but with additional germanium filter; 3) decay curve.

positions of the two maxima; transmission outside the pass band decreased from 10^{-2} to 10^{-4}. This was very important at long wavelengths far from the emission maximum of the incandescent lamp. The presence of even very small numbers of transmitted high-energy quanta would have distorted the results very considerably. For example, a flash was observed (curve 1 in Fig. 3) under the influence of infrared radiation transmitted by a set of filters with transmission maxima at 2.08 and 2.13 μ. When this set was supplemented by a germanium filter, which removed all wavelengths shorter than 1.74 μ and transmitted about 50% of the light energy in the region of 2 μ, the flash disappeared completely (curve 2) although the transmission of the 2.08 μ + 2.13 μ + Ge filter combination was over 50% higher than the transmission of the 208 μ + 2.13 μ combination. Therefore, in all cases it was desirable to use suitable germanium or silicon cutoff filters.

The nominal number of infrared photons reaching a phosphor was determined with a thermocouple fitted into the cryostat at the point where the investigated phosphor was located. A measure of the number of these quanta was provided by the ratio of the thermo-emf produced by the illuminated thermocouple to the value of h ν corresponding to the transmission maximum of the filter combination employed. Attenuating grids were used to ensure that in all cases the number of photons transmitted by different filter combinations did not differ by more than one order of magnitude.

The brightness of the afterglow and flash were recorded with a detector —amplifier system comprising an FÉU-19 M photomultiplier, a dc amplifier, and a recorder ultilizing photosensitive paper. The spectral range of the sensitivity of the FÉU-19 M photomultiplier was 3500-6000 Å. The maximum sensitivity of this photomultiplier was 100 A/lm when the applied voltage was 1200 V. The dark current of the photomultiplier was 10^{-7} A. The photomultiplier was supplied with a stabilized voltage of 1-1.5 kV. It was screened by a cell containing a solution of $CuSO_4$ for protection from reflected infrared radiation and by a cell with a solution of $NaNO_2$ for protection from reflected ultraviolet light; if necessary, use was made also of a yellow filter transmitting light of $\lambda > 0.5$ μ wavelength or a blue filter with a pass band from 0.3 to 0.5 μ when selection of the green or blue luminescence of ZnS:Cu:Co:Cl was required. A check was made to make sure that the reflected infrared radiation and the stray scattered light gave rise to a signal not exceeding two divisions on the most sensitive photomultiplier scale. Thereafter the influence of such reflected and stray light could be ignored.

The signal produced by the FÉU-19 M photomultiplier was applied to a dc amplifier consisting of a bridge circuit based on two 6Zh1Zh pentodes operated in the electrometric regime. The amplification factor varied from 10^3 to 10^6, depending on the scale employed. The dark current was compensated by a reverse current. Moreover, compensation was sometimes used to remove the constant component of the signal so that the flash could be measured using the most sensitive scale. This amplifier was supplied from a stabilized transistor rectifier. A

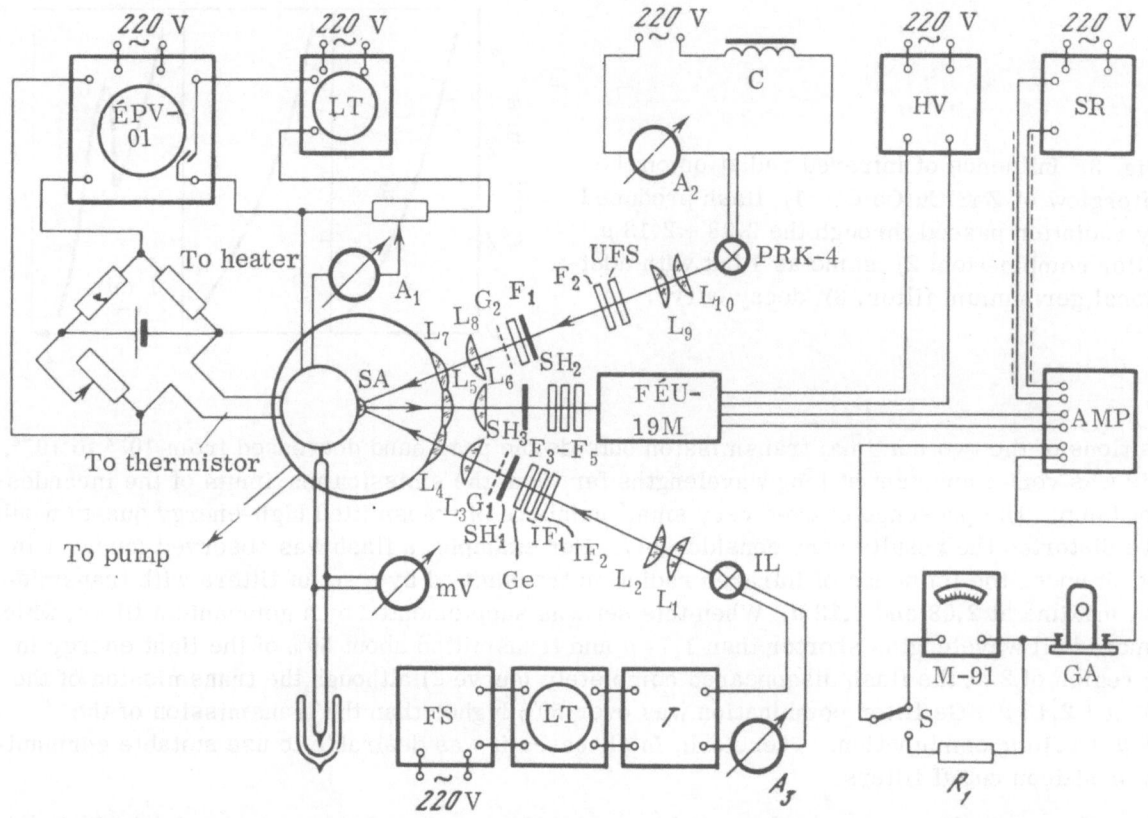

Fig. 4. Block diagram of the apparatus. Here, A_1-A_3 are ammeters; L_1-L_{10} are lenses; SH_1 and SH_2 are shutters; G_1 and G_2 are attenuating grids; F_1-F_5, Ge, IF_1, IF_2, and UFS are filters; IL is an incandescent lamp; HV is a high-voltage source; AMP is a dc amplifier; SR is a stabilized rectifier; GA is a short-period galvanometer; C is a choke; T is a current transformer; LT is a laboratory transformer; FS is a ferroresonance stabilizer; SA is a sample.

special check was made that the whole measuring system responded linearly to the illumination intensity. The amplified current was measured with an M-91 galvanometer of 10^{-7} A/division sensitivity and it was recorded on photosensitive paper using a short-period galvanometer (Fig. 4). A switch S made it possible to disconnect the M-91 galvanometer from the line supply, replacing it with an equivalent resistance R_1; this enabled us to record rapidly growing flashes.

The absolute sensitivity of the system, measured with a constant-intensity luminescence source, was 10^{10} photons $\cdot$ sec$^{-1} \cdot$ cm$^{-2} \cdot$ division^{-1}. Thus, the system was capable of detecting changes in the phosphor luminescence brightness not smaller than 10^{10} photons $\cdot$ sec$^{-1} \cdot$ cm^{-2}. The temperature of a sample was kept constant with a controller, consisting of an ÉPV-01 automatic potentiometer and a rheostat. The winding of a thermistor was connected as one of the arms of a bridge. The unbalance voltage was picked up from the bridge diagonal and applied to the ÉPV-01 potentiometer. This method of connecting ÉPV-01 considerably increased its sensitivity and made it possible to maintain the temperature of a sample constant to within tenths of a degree for a long time.

A coarse temperature adjustment was made by altering the position of the liquid nitrogen reservoir and the current in the heater. A fine adjustment was provided by moving the bridge slider and the contact in the ÉPV-01 potentiometer. The bridge was balanced when the thermis-

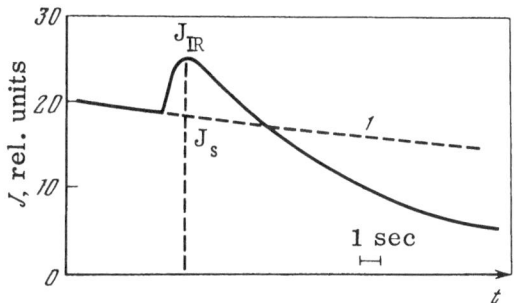

Fig. 5. Calculation of the flash amplitude J_{IR} − J_s. Curve 1 represents undisturbed decay (afterglow).

tor was at the required temperature. Then, the ÉPV-01 potentiometer switched off the heater current. The cold part of the cryostat lowered gradually the temperature of the sample and reduced the thermistor resistance. This unbalanced the bridge, and the ÉPV-01 potentiometer switched on the heater until equilibrium conditions were reestablished. In this way the temperature of the sample fluctuated about the required value within a few tenths of a degree, and the frequency and amplitude of these fluctuations depended on the coarse-adjustment regime. The temperature was measured with a copper − Constantan thermocouple.

In measurements of stimulated flashes the infrared illumination was started at some definite afterglow stage when the brightness fell to some selected value J_s. The value of J_{IR} was the brightness at the flash maximum. When the natural decay was so fast that during the flash rise time the afterglow brightness decreased significantly, a necessary correction was made to the value of J_s (Fig. 5). Since the behavior of the curves and not the absolute values of the coordinates was important in the determination of E_{opt} and E_{th}, we took I to be the nominal number of photons as defined above and we replaced R with the difference $J_{IR} - J_s$, because in each series the value of J_s remained constant. The measurements were repeated three times to obtain one experimental point.

§ 4. Investigated Samples and Excitation Method

Zinc sulfide is characterized by a considerable proportion of ionic binding and, therefore, the difference between the optical and thermal depths of traps in ZnS crystals should be large. According to [1], this difference is 0.55 ± 0.15 eV. This value was obtained in a study of deep traps in ZnS:Cu:Co:Cl crystal phosphors. Therefore, we studied shallow traps in the same phosphors. These phosphors stored a large light sum and produced strong flashes when subjected to infrared radiation. All samples were heated in air under an NaCl flux (5%) at 1100°C for 30 min. The measurements were carried out mainly on the "optimal" phosphor composed of ZnS, 10^5 g/g Cu, 10^{-6} g/g Co, and Cl; we used also other concentrations of Cu and Co lying within the range 10^{-5}-10^{-6} g/g. Thermoluminescence of several phosphors of this type was studied in [5]. It was then found that there were at least five thermoluminescence peaks at −130, −60, 0, 20, and 80°C. They were attributed to excess zinc, oxygen, copper (both 0 and 20°C), and cobalt, respectively. The thermal depth of the cobalt level was measured by different methods which gave 0.42 eV [4], 0.44 eV [6], and 0.40 eV [2]. The optical depths of the other traps have not yet been measured sufficiently reliably.

We selected the exciting light intensity so that the fraction of the total light sum stored in the shallow traps was as high as possible. This was done on the basis of the results of a study of the dependence of the steady-state luminescence brightness J_{st} on the excitation intensity E and by plotting the dependence of log (J_{st}/E) on log E.

When the phosphor layer was sufficiently thick so that the exciting light was absorbed completely, the ratio J_{st}/E was proportional to the energy efficiency of the luminescence. At high excitation intensities this efficiency began to fall when the excitation was increased fur-

ther, because some of the exciting light quanta were absorbed by electrons in traps causing
liberation which could have occurred otherwise under the influence of thermal vibrations.
The fall of the efficiency was accompanied by a rise in the density of electrons localized in
traps [7]. This occurred for the same reason: An increase in the excitation intensity increased
not only the supply of electrons to the traps, but also liberated some of these electrons.

When the excitation intensity was such that the probability of optical liberation of electrons from given traps was greater than the probability of their thermal liberation, the density
of electrons localized in these traps ceased to depend on the excitation intensity. If the shallower traps then contained fewer localized electrons, the total number of localized electrons
was almost constant in a certain range of excitation intensities. Therefore, in the same range
the useless absorption of light and the luminescence efficiency should also be constant. At
higher excitation intensities the total number of localized electrons should again increase but
this time due to an increase in the density of electrons in the shallow traps. The efficiency
should start to fall again. In other words, in this range of values of E there should be a step
in the dependence $\log (J_{st}/E) = f(E)$. This step should shift toward higher excitation intensities
when the temperature was increased because this increased the probability of thermal liberation of electrons.

In studies of shallow traps one should use the highest possible excitation intensity because this would concentrate the greatest proportion of electrons in the shallow traps. However, at all temperatures used in the measurements the concentration of ionized luminescence
centers and, consequently, the total density of localized electrons should be the same. This
would ensure that the conditions in the phosphor were comparable at different temperatures.
Therefore, it would be undesirable to go outside the region of this step.

We found (Fig. 6) that the relative efficiency J_{st}/E did indeed fall with rising E, and this
fall was nonmonotonic: In the range of high values of E the curve had a step, i.e., an almost
horizontal region which was followed by a steep fall. Following the considerations given above,
the intensity of the exciting light was selected so that it corresponded approximately to the
midpoint of this step.

We selected natural diamonds as a sample with the covalent binding which should not
exhibit a difference between the optical and thermal trap depths. Diamond samples were excited with ultraviolet light of $\lambda \approx 300$ nm wavelengh selected from the spectrum of the PRK-4
lamp with a UFS-1 filter and $NiSO_4$ aqueous solution.

Out of three investigated diamond crystals only two were sensitive to infrared radiation.
The third crystal exhibited a bright luminescence during excitation but was insensitive to infrared radiation probably because the light sum stored in this crystal was close to zero (this
was indicated by the absence of thermoluminescence). Crystal A_3 had a single large thermolu-

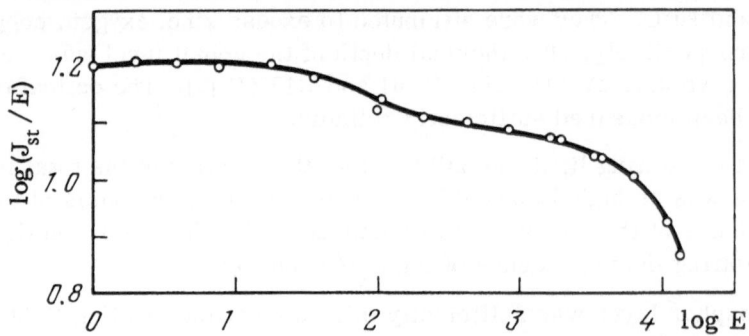

Fig. 6. Dependence of $\log (J_{st}/E)$ on $\log E$ for ZnS:Cu:Co:Cl.

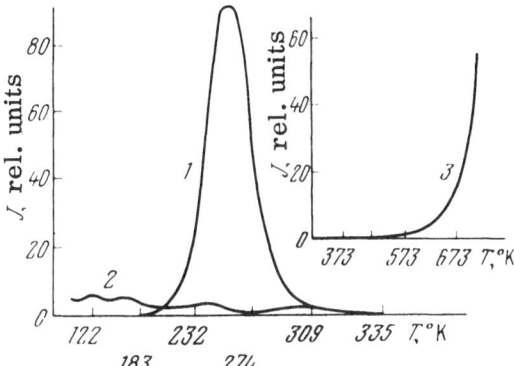

Fig. 7. Thermoluminescence curves of two diamond samples: 1) A3; 2) A4 (ordinate scale increased by a factor of 10); 3) A4 recorded at a higher rate of heating.

minescence peak at 255°K (curve 1 in Fig. 7). This crystal produced a flash under the influence of infrared radiation only at low temperatures. Crystal A4 produced a flash also at room temperature. It had two small thermoluminescence peaks at 122 and 165°K and a very large peak above 400°K.

§ 5. Optical Depths of Traps in ZnS

An investigation of the ZnS:Cu:Co:Cl phosphor at room temperature indicated that when $h\nu$ was increased right up to 0.643 eV (using filters with a transmission maximum at 1930 nm), a flash became practically invisible. It appeared beginning from $h\nu$ = 0.67 eV (when filters with a transmission maximum at 1850 nm were used). Figure 8 shows the dependences $\ln(R/I_{h\nu})$ = $f(h\nu/kT)$ consisting of two parts with very different slopes. The part with the less steep slope clearly corresponded to $h\nu > E_{opt}$ for the traps under investigation. This slope was practically unaffected when the beginning of infrared illumination was delayed from 30 sec after the end of excitation to 2 and 8 min. In the spectral range 0.689 eV $\leq h\nu \leq$ 0.67 eV, i.e., between 1800 and 1850 nm, the flash amplitude fell so rapidly that the slope of the curve even exceeded unity. This could not be due to some unidentified experimental error because the slope exceeding unity was observed outside the stated spectral range. In fact, a flash was still observable for $h\nu$ = 0.67 eV. We then used a filter combination with a transmission maximum at $h\nu$ = 0.643 eV. As pointed out earlier, a flash was not observed for this combination. Allowance for the transmission of the filters and the capabilities of the recording system indicated that the sensitivity to infrared radiation fell in this region by a factor of at least 2-3. Therefore, even when it was assumed that the last point was incorrect and it was excluded from consideration and if it was replaced with the point at $h\nu$ = 0.643 eV, postulating a small

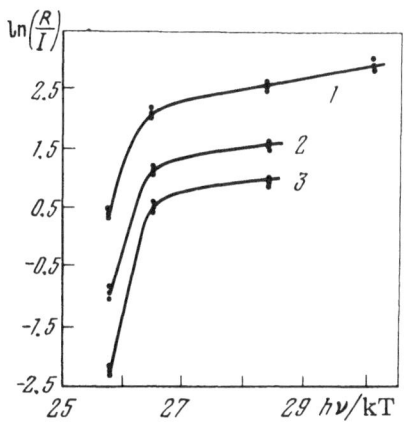

Fig. 8. Dependence of $\ln(R/I_{h\nu})$ on $h\nu/kT$ for ZnS:Cu:Co:Cl at room temperature; 1) infrared illumination started 30 sec after the end of excitation; 2) after 2 min; 3) after 8 min.

flash amplitude, the steep part of the curve still had a slope exceeding unity. Therefore, this slope should be regarded as firmly established. It started somewhere near $h\nu = 0.69$ eV. Consequently, the optical depth of the investigated traps was 0.69 ± 0.02 eV.

The error of 0.02 eV was approximately equal to the difference between the values of $h\nu$ at two successive points where the measurements were made, but even if the points were more frequent one could not regard the results as more accurate. This was due to the fact that, in principle, the optical (and even the thermal) depth of a trap could hardly be determined with an error much smaller than kT, which was 0.026 eV at room temperature.

At 100°K the flash sensitivity spectrum was measured between 1500 and 4000 nm. It was found that the sensitivity fell nonmonotonically with increasing wavelength: In the region of 2300 nm (0.55 eV) there was a small maximum which was observed also for other zinc sulfide phosphors activated with various elements (see, for example, [8-10]). In our case this maximum appeared only when the flash sensitivity spectrum was recorded during the early decay stages but was not observed during the later stages.

There was also a small flash maximum in the region of 3700 nm (0.34 eV) and this maximum appeared at any decay stage. However, we were unable to study it in greater detail because of its low amplitude.

Figure 9 shows the flash sensitivity spectra of the investigated phosphor at 100°K plotted using the coordinates $\ln\left[(J_{IR} - J_s)/I\right]$ and $h\nu/kT$. These spectra exhibited clearly two regions of rapid fall: One was located at $h\nu/kT \approx 45$ ($h\nu \approx 0.38$ eV) and the other at $h\nu/kT \approx 75$ ($h\nu \approx 0.60$-0.70 eV). The second region consisted of two parts separated by a interval with a gentle slope. This indicated a possible existence of two systems of traps with similar optical depths. It was found that a system of traps with the greater optical depth was identical with that whose optical depth had already been determined at room temperature.

The slope of the curves in the $h\nu/kT \approx 45$ region was close to unity, whereas in the region $h\nu/kT \approx 75$ it was much less than unity, but during later stages of the afterglow it increased, approaching unity. This occurred because during decay the density of electrons in shallow traps decreased more rapidly than in deep traps and electrons were transferred from shallow to deep traps. This explained also the gradual disappearance of the step in the region of 0.45-0.50 eV. The existence of systems of traps in the region $h\nu \approx 0.63$-0.69 eV became manifest when corrections were made for the liberation of electrons from shallow traps.

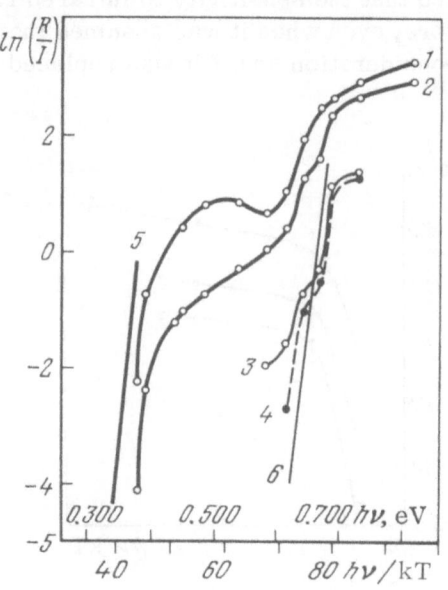

Fig. 9. Dependence of $\ln (R/I_{h\nu})$ on $h\nu/kT$ for ZnS:Cu:Co:Cl at 100°K: 1) infrared illumination started 30 sec after the end of excitation; 2) after 8 min; 3) after 20 min; 4) curve 3 after correction for the flash due to shallow traps; 5, 6) lines with slopes equal to unity.

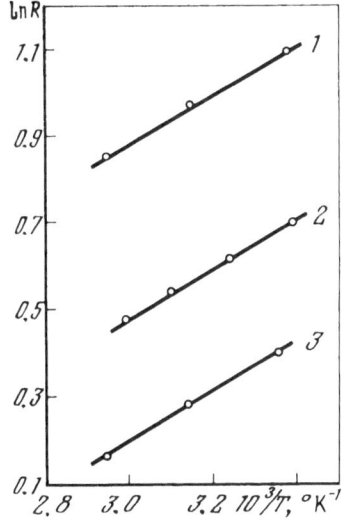

Fig. 10. Dependence of ln R on 1/T for ZnS:Cu:Co:Cl: 1) infrared illumination started 5 min after the end of excitation; 2) 8 min after; 3) 11 min after.

Thus, at low temperatures we were able to observe at least three systems of traps with optical depths at $E_1 = 0.38 \pm 0.01$ eV, $E_2 = 0.63 \pm 0.02$ eV, and $E_3 = 0.69 \pm 0.02$ eV. The last trap system was also observed at room temperature. The experimental error did not exceed the width of the region of linear fall lying between the two flat regions. A system of traps with an optical depth of ≤ 0.34 eV was also likely to be present.

§ 6. Thermal Depths of Traps in ZnS

The method described above was used to determine the thermal depth of traps which were observed at room temperature and whose optical depth was 0.69 ± 0.02 eV. The flash amplitude was measured between room temperature and +60°C during different stages of the decay. A flash was produced by a set of filters with a transmission maximum at 1800 nm ($h\nu \approx 0.69$ eV). The results were plotted in Fig. 10 using the coordinates $\ln[(J_{IR} - J_s)/J_s]$ and $10^3/T$.

It is clear from Fig. 10 that points obtained during one afterglow stage fitted well straight lines which were parallel to one another. The common slope of these lines was used to find the thermal depth of traps $E_{1th} = 0.06 \pm 0.01$ eV.

It was interesting to note that this small depth was obtained at room temperature during relatively late stages of the afterglow (5-11 min after the end of excitation) when almost all the electrons were concentrated in the deeper traps. This was in full agreement with the theory set out in § 1.

This theory predicted also that the value of R should be independent of the luminescence band whose flash stimulation was measured. This prediction was checked experimentally. The temperature dependence of the flash amplitude obtained for ZnS:Cu:Co:Cl illuminated with infrared radiation of wavelength 2580 nm (0.48 eV) at temperatures between 90 and 150°K was plotted (Fig. 11) using the coordinates $\ln[(J_{IR} - J_s)/J_s]$ and $10^3/T$. We determined the integrated brightness of the flash (curve 3a in Fig. 11), as well as the separate brightness of the green (curve 3b) and blue (curve 3c) bands during the same afterglow stage.* We found that all

---

* The corresponding optical depth of the traps was 0.38 eV. However, in view of the smallness of the amplitude of the flash produced by the radiation selected with a set of filters having a maximum at $h\nu = 0.38$ eV, we used sets of filters with $h\nu = 0.44$ and 0.48 eV to measure the thermal depth. The results obtained by means of the last two filter sets were identical.

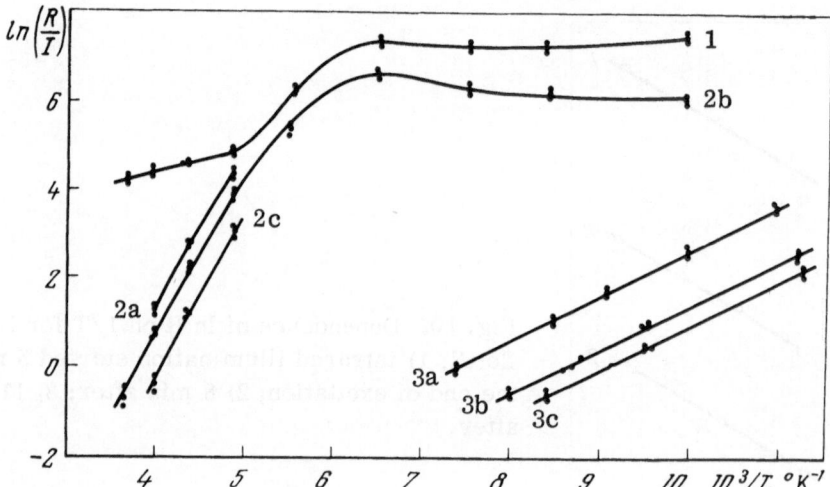

Fig. 11. Dependence of ln $(R/I_{h\nu})$ on $1/T$ for ZnS:Cu:Co:Cl: 1)
flash produced by radiation of $h\nu$ = 0.69 eV energy; 2) $h\nu$ = 0.63
eV (curves 2a, b, and c correspond to decay stages when J_s =
400, 200, and 100 rel. units); 3) $h\nu$ = 0.48 eV (curves 3a, b, and c
represent integrated, green, and blue luminescence flash amplitudes obtained during the same decay stage).

three curves gave identical results: The points belonging to the same band fitted well straight
lines which had almost the same slope. This slope yielded the thermal depth of 0.08 ± 0.02 eV.

　The following advantage of the selected method over the thermoluminescence technique
was manifested in these measurements. When the thermoluminescence curves were recorded,
the ratio of the intensities of the green and blue bands varied continuously. Therefore, corrections were necessary when the luminescence measurements were made with a photomultiplier or
some other detector whose sensitivity varied with the wavelength. In the method employed in the
present study this was not necessary because the final result did not include the luminescence
brightness.

　Figure 11 shows also the temperature dependence of the flash amplitude produced by
illumination with wavelength 1970 nm ($h\nu$ = 0.63 eV) during three different stages of the afterglow (curves 2a-c). It is clear from the figure that all of them had parallel linear regions, in
agreement with the theory. The thermal depth of the traps deduced from the slopes of these
regions was 0.34 ± 0.03 eV. This (or similar) depth was also found for ZnS:Cu:Cl phosphors
which did not contain Co [2, 6, 11, 12]. Therefore, it was not surprising that the same trapping
level was also found in our ZnS:Cu:Co:Cl phosphor.

　It is also clear from Fig. 11 that the temperature dependence of the flash amplitude was
complex. As pointed out in [2], in the case of the ZnS:Cu:Cl phosphor this dependence had two
linear regions with different slopes and the flatter of these regions was observed at lower
temperatures than the steeper part. This was explained as manifestation of two types of trap
(0.3 and 1.1 eV) and it seemed natural that the shallower traps were observed at lower temperatures. However, it is clear from curve 1 in Fig. 11 that the opposite could also be true when
a steeper part was located on the side corresponding to lower temperatures. A comparison
of curves 1 and 2b readily demonstrated that at low temperatures they were almost parallel
whereas at high temperatures (on the left-hand side of Fig. 11) their slopes differed considerably. In this region the flatter part corresponding to the shallower trap was the curve describing a flash produced by photons of higher energy. This observation, as well as the "incorrect"

positions of the flatter and steeper parts, can be explained using the energy band scheme considered in § 1. Let us assume that the equilibrium conditions given by Eq. (10) are satisfied by all three traps in the investigated phosphors (their depths are 0.06-0.08 eV, 0.34 eV, and − according to [2] − 0.40-0.44 eV). Then, following Eq. (9), we obtain

$$\frac{w_1 n_1}{\sigma_1 v_1} = \frac{w_2 n_2}{\sigma_2 v_2} = \frac{w_3 n_2}{\sigma_3 v_3}.$$ (15)

If the other traps are so deep that the liberation of electrons from them can be ignored, Eq. (5) for R can be expressed three ways in terms of n_1, n_2, and n_3 bearing in mind which term predominates in the numerator. This means that the temperature dependence of the flash amplitude gives the depth of those traps from which most of the electrons are liberated by given infrared radiation.

When $h\nu$ exceeds the optical depth of the deepest traps, the numerator of Eq. (5) should be dominated by the last term provided ν_3 is not too small. Then, the temperature dependence of R is governed by the value of w_{3th}:

$$R = \frac{p_3(h\nu) I}{w_{3th}} \left(\frac{\sigma_1 v_1}{\sigma_3 v_3} + \frac{\sigma_2 v_2}{\sigma_3 v_3} + 1 \right)^{-1}.$$ (16)

In the case of ZnS:Cu:Co:Cl this corresponds to $h\nu \geq 1$ eV and a linear region of slope 0.4 eV. However, if $h\nu$ exceeds the optical depth of the shallowest traps, we can ignore the first term in Eq. (5). In this case, instead of Eq. (16), we obtain a formula for R in which the quantities describing shallow and deep traps are interchanged. In our case this corresponds to $h\nu = 0.69$ eV at temperatures above 220°K.

We shall now find the conditions under which the second term predominates in Eq. (5). As long as $h\nu$ exceeds the optical depth of the shallowest traps, approaching the optical depth E_2 of the intermediate traps ($h\nu \leq E_2$), the quantity $p_1(h\nu)$ does not change greatly when the temperature is varied, whereas $p_2(h\nu)$ decreases when the temperature is lowered because of a reduction in the probability of acquiring the additional thermal energy needed for the liberation of electrons from these traps if electron transitions occur in accordance with the Franck−Condon principle. If this principle is not obeyed, the liberation of electrons requires less energy because ions in the crystal lattice surrounding the trap under consideration assume immediately a new equilibrium state. Although the probability of such electron liberation is much lower than the probability of liberation subject to the Franck−Condon principle, it varies only weakly with temperature, and in a certain spectral range it is also practically independent of the photon energy. Therefore, if an excessive thermal energy is required for the liberation of an electron in the "lawful manner" (in accordance with the Franck−Condon principle), liberation violating the Franck−Condon principle may become more probable. In this case the temperature dependence of ratio of the first and second terms in the numerator of Eq. (5) is governed by the change in the ratio n_1/n_2, which can be derived from Eq. (15):

$$\frac{n_1}{n_2} = \frac{v_1}{v_2} \exp\left[-\frac{E_{2th} - E_{1th}}{k T} \right].$$ (17)

When the difference $E_{2th} - E_{1th}$ is large, the above ratio decreases rapidly with increasing temperature. Therefore, beginning from a certain temperature we may find that $p_1 n_1 \ll p_2 n_2$, in spite of the fact that $p_1 \gg p_2$. This produces a kink in curve 1 of Fig. 11 at T < 220°K.

When $h\nu$ is less than the optical depth of the shallowest traps, this predominance is even greater because at these values of $h\nu$ the value of $p_1(h\nu)$ varies rapidly, whereas $p_2(h\nu)$ varies slowly. It is clear from Fig. 11 (curves 2a-c) that in the case of $h\nu = 0.63$ eV the region with the steeper slope extends to at least 250°K. On the other hand, the flash produced by $h\nu = 0.63$ eV is one or two orders of magnitude weaker than that produced by $h\nu = 0.69$ eV, which shows that $p_1(h\nu)$ decreases.

For even lower values of $h\nu$ the photon energy is insufficient for the liberation of electrons from the deepest traps even when the Franck−Condon principle is violated. Once again the liberation from the shallow traps begins to predominate but now subject to the violation of this principle. In our experiments this is manifested by the fact that illumination with $h\nu = 0.48$ 0.48 eV again produces a region with a gentle slope (0.08 eV), as manifested by curves 3a−c in Fig. 11.

Thus, the temperature dependence of the flash amplitude at $T > 220°K$ demonstrates not only a redistribution of electrons between the traps, but also that the optical liberation of electrons may occur when the Franck−Condon principle is violated, and then photons of lower energy are sufficient.

§ 7. Adiabatic and Nonadiabatic

Liberation of Electrons from Traps in ZnS

The results of measurements of the optical and thermal depths of traps of various kinds in ZnS are collected in Table 1.

It is clear from this table that, within the limits of the experimental error, the difference between the optical and thermal depths is either 0.6 or 0.3 eV and this applies to shallow as well as to deep traps.

The optical depth of the 0.38-eV traps and the difference between the optical and thermal depths of these traps are less than twice the binding energy of polarons in ZnS, given in [1] as 0.5-0.6 eV. This may be because an electron trap of this kind describes a large-radius hydrogen-like orbit and, therefore, produces a weak polarization of the lattice. In this case there is hardly any relaxation of the crystal lattice after the removal of an electron from a trap of this kind, and the binding energy of a polaron is evolved only once: when a photoliberated electron is converted into a polaron. Therefore, the total relaxation energy of the lattice (and the difference between the optical and thermal trap depths equal to this energy) should be only slightly larger than the binding energy of one polaron.

The difference between the optical and thermal depths of the traps in ZnS may be small also when an electron is localized in a small volume but the optical liberation occurs in such a way that the Franck−Condon principle is violated. In this case the transition probability is

TABLE 1

E_{th}, eV	$E_{h\nu}$, eV	$(E_{h\nu} - E_{th})$, eV
0.06 ± 0.01	0.69 ± 0.02	0.63 ± 0.02
0.08 ± 0.02	0.38 ± 0.01	0.30 ± 0.03
0.34 ± 0.02	0.63 ± 0.02	0.29 ± 0.04
0.44 *		
0.56 *		0.55 ± 0.15 †

* These results were taken from [2] for the cobalt and nickel levels in ZnS:Cu:Co:Cl and ZnS:Cu:Ni:Cl.

† These results were obtained using infrared radiation that was not fully monochromatic so that the exact optical depth of traps could not be determined.

very low. Nevertheless, as soon as hν becomes smaller than the sum of the thermal trap depth plus the binding energy of one polaron, this probability decreases exponentially on further reduction of hν since the liberation of an electron requires also an additional thermal energy.

However, the alternate manifestation of the shallow and deep traps observed when the temperature of a sample is increased gradually can be explained only by the existence of electron liberation processes characterized by low and high probabilities. The large difference between the optical and thermal depths of the shallow traps and the small difference between the depths of the deep traps is in conflict with the assumption that the small difference between these depths is due to the liberation of electrons moving along large-radius hydrogen-like orbits. It follows from the foregoing discussion that the capture of electrons by deep and shallow traps in ZnS causes ionic polarization of the lattice, i.e., it displaces neighboring ions from their equilibrium positions, and the optical liberation of electrons may occur in two ways, one of which is adiabatic and the other nonadiabatic.

As mentioned earlier, in an adiabatic optical liberation event an electron transition occurs so rapidly that, in accordance with the Franck−Condon principle, the surrounding ions are not displaced and their vibrational velocity is not affected. Therefore, after liberation of an electron these ions are in a nonequilibrium position. The energy released as a result of the transition of these ions to new equilibrium positions should naturally be transferred to the trap by the same photon which has caused the electron liberation. However, this does not provide all the additional energy needed for the optical liberation of electrons. This is due to the fact that a liberated electron finally polarizes the lattice and becomes a polaron. This evolves an energy which is approximately the same as the energy evolved in a trap after the removal of an electron. On the other hand, the thermal liberation process is of quasiequilibrium type. Immediately before and after this process an electron and a hole are in an equilibrium with the surrounding medium. Thus, in accordance with the Franck−Condon principle, the optical depth of a trap under adiabatic liberation conditions should be greater than the thermal depth by an amount approximately equal to twice the binding energy of a polaron.

The optical liberation of electrons from traps may also occur nonadiabatically, and in this case a photon of lower energy is sufficient because the transition of the surrounding ions to a new equilibrium position occurs at the same time as the removal of an electron from a trap. Therefore, after the liberation of an electron the lattice surrounding a trap does not relax until the liberated electron is converted into a polaron. Therefore, the minimum photon energy sufficient for a nonadiabatic liberation of an electron should be less than the minimum photon energy in the adiabatic case, and the difference between these two energies should be approximately equal to the binding energy of one polaron.

In the case of ZnS:Cu:Co:Cl the thermal depth of 0.06 eV obtained close to room temperature agrees, within the limits of the experimental error, with the depth 0.08 eV found at lower temperatures. It follows that the traps in question have two optical depths: The greater depth corresponds to the adiabatic liberation of an electron and the smaller depth, to the nonadiabatic process. The optical depth of the traps characterized by E_{th} = 0.34 eV (Table 1) clearly corresponds to the nonadiabatic liberation of electrons, and the difference between the optical and thermal depths of traps with E_{th} = 0.44 and 0.56 eV corresponds to the adiabatic process. It is also clear from Table 1 that the difference between the adiabatic and nonadiabatic optical depths of traps is equal to the difference between the nonadiabatic and thermal depths.

Having measured the amplitude of a flash under the influence of photons causing adiabatic and nonadiabatic liberation of electrons from the same traps, we can estimate the ratio of the probabilities of these processes. In fact, if infrared radiation of energy $h\nu_1$ is capable of liberating electrons from shallow traps E_1 both adiabatically and nonadiabatically but only nonadiabatically from intermediate traps E_2, the relative amplitude R' of a flash is of the form*

$$R_1' = \frac{J_{\text{IR}} - J_s}{I_1} = \text{const} \left[(p_1^*(h\nu_1) + p_1(h\nu_1)) n_1 + p_2^*(h\nu_1) n_2 \right]. \tag{18}$$

Here, $p_1^*(h\nu_1)$ and $p_2^*(h\nu_1)$ are the coefficients of the nonadiabatic optical liberation of electrons from the traps E_1 and E_2; $p_1(h\nu_1)$ is the coefficient of the adiabatic optical liberation (in our specific case we have $h\nu = 0.69$ eV); I_1 is the infrared radiation intensity expressed as the number of photons. If infrared radiation can liberate electrons from the traps E_1 and E_2 only nonadiabatically, we find that

$$R_2 = \frac{J_{\text{IR}} - J_s}{I_2} = \text{const} \left[p_1^*(h\nu_2) n_1 + p_2^*(h\nu_2) n_2 \right], \tag{19}$$

where $h\nu_2 = 0.63$ eV. Finally, if infrared radiation can liberate electrons only nonadiabatically from the shallow traps ($h\nu_3 \approx 0.48$ eV), we find that

$$R_3 = \frac{J_{\text{IR}} - J_s}{I_3} = \text{const} \left[p_1^*(h\nu_3) n_1 \right]. \tag{20}$$

If R_1, R_2, and R_3 are measured under identical excitation conditions, at the same temperatures, and during the same decay stage, the values of the constants in front of the brackets in Eqs. (19)–(20) are the same and we then find

$$\frac{p_1(h\nu_1)}{p_1^*(h\nu_3)} = \frac{R_1 - R_2}{R_3}. \tag{21}$$

This formula is derived on the assumption that $p_1^*(h\nu_1) = p_1^*(h\nu_2)$ and $p_2^*(h\nu_1) = p_2^*(h\nu_2)$. The formula (21) underestimates somewhat the probability ratio because in reality the quantity R_2 includes a contribution due to electrons liberated adiabatically by infrared radiation of $h\nu_2 = 0.63$ eV energy from the shallow traps with an addition of the thermal energy. This contribution is ignored in Eq. (19). The results calculated using Eq. (21) demonstrate that the probability of the nonadiabatic liberation of an electron is two or three orders of magnitude lower than the probability of the adiabatic liberation. As expected, the nonadiabatic transition is forbidden. Hence, the spectral region where the nonadiabatic liberation predominates may exist only if the polaron binding energy is 5–7 times greater than kT. In the opposite case the probability of acquisition of an additional thermal energy needed for the adiabatic liberation of an electron is always greater than the probability of the nonadiabatic process. Since the polaron binding energy is 0.3 eV for ZnS, this condition is satisfied even at room temperature. However, in the case of substances which are less ionic so that the polaron binding energy is less, a considerable cooling of the sample may be necessary for the observation of the nonadiabatic liberation of electrons.

§8. Optical and Thermal Depths of

Traps in Diamond

The results of measurements of the optical depth of traps in diamond samples A3 and A4 are plotted in Figs. 12 and 13. The dependence of $\ln \left[(J_{\text{IR}} - J_s)/I \right]$ on $h\nu/kT$ for sample A3 is simple. The horizontal part of the dependence represents transitions induced by photons of energy exceeding the optical trap depth whereas the linear region represents transitions with $h\nu \sim E_{h\nu}$. We can see that this crystal contains traps of just one kind and the optical depth of

---

* The quantity R' is related to R by R' = RJ/I.

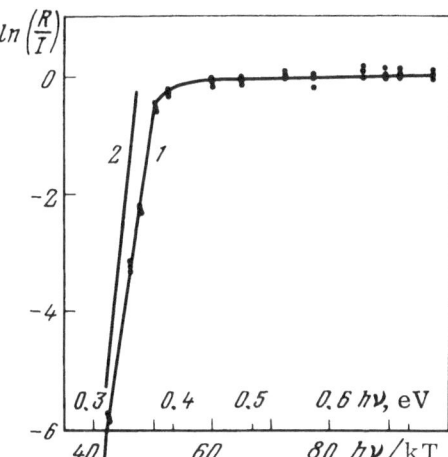

Fig. 12. Determination of the optical depth of traps in diamond sample A3: 1) dependence of ln $(R/I_{h\nu})$ on $h\nu/kT$ at 87°K; 2) straight line of slope equal to unity.

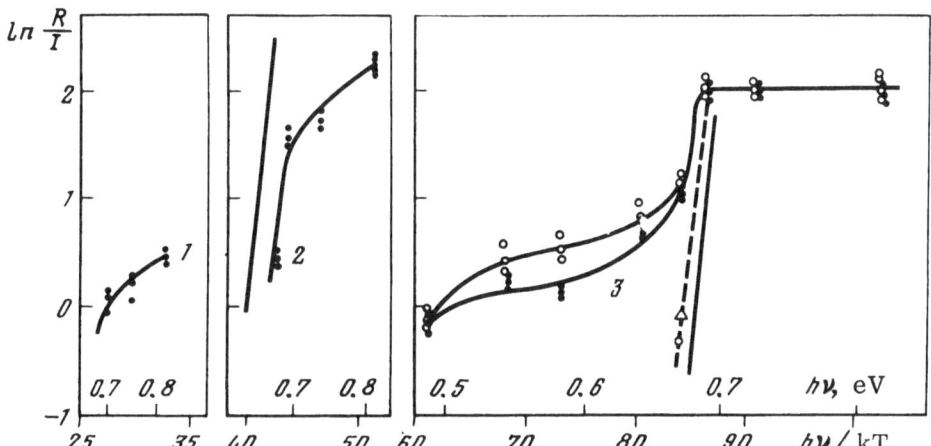

Fig. 13. Determination of the optical depth of traps in diamond sample A4: 1) dependence of ln $(R/I_{h\nu})$ on $h\nu/kT$ at room temperature; 2) at 183°K; 3) at 92°K (upper curve for J_s = 15 rel. units and lower curve for J_s = 2 rel. units); the continuous lines have a slope of unity; the dashed line represents both experimental curves after correction for the flash due to shallow traps.

these traps is $E_{h\nu}$= 0.32 ± 0.02 eV. This depth is in agreement with the value obtained by many workers for class IIb diamonds from the infrared impurity absorption spectra (see, for example, [13-15]).

It is clear from Fig. 13 that sample A4 has at least two types of trap: Shallow traps are responsible for the lower horizontal part of the curve and deep traps responsible for the upper horizontal part. During later stages of the decay (afterglow) the long-wavelength horizontal part of the curve drops considerably but the short-wavelength part is not affected. Moreover, it is clear from curve 2 that at 183°K, which lies above two small thermoluminescence peaks, the whole long-wavelength part is cut off and the curve is reduced to the part which has a fall in the region of 0.67 eV with a slope close to unity. After correction for the amplitude of the flash caused by the liberation of electrons from shallower traps, the fall of curve 3 in the region of 0.67 eV also becomes linear and the slope is about unity. All this indicates that sample

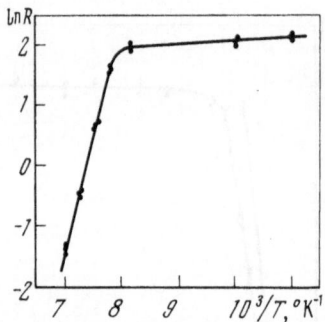

Fig. 14. Dependence of ln R on 1/T for diamond sample A3.

A4 has traps with a depth of 0.67± 0.02 eV. This depth also agrees with the published work [16, 17]. The depth of the shallower traps could not be measured because of the insufficient sensitivity of the apparatus.

The thermal depths of the traps whose optical depths were $E_{h\nu}$ = 0.32 and 0.67 eV, and which were found in samples A3 and A4, were measured using infrared radiation of energy $h\nu$ = 0.38 and 0.81 eV, respectively. The plotted function $\ln[(J_{IR} - J_s)/J_s] = f(1/T)$ had linear regions in the temperature range from 127 to 142°K for sample A3 (Fig. 14) and in the range from 274 to 314°K for sample A4 (Fig. 15). The slopes of these linear regions yielded the thermal depths 0.33 ± 0.01 and 0.66 ± 0.02 eV, respectively. Clearly, within the limits of the experimental error, the optical and thermal depths of the same system of traps in diamond were identical.

The thermal depth of about 0.34-0.35 eV obtained for diamond was also reported by other workers who measured the temperature dependences of the Hall coefficient or resistivity (see also [13]). Traps of this depth were found to be acceptors responsible for the p-type conduction of the majority of diamonds. There was no definite evidence for identifying the traps found in our study with the acceptors reported by other workers, but it was quite likely that in our case the flash was also due to hole transitions. This did not affect the conclusions drawn from our results because of the complete symmetry between electrons and holes in our study.

Numerous data are available on the optical and thermal depths of the same traps in several other covalent crystals. For example, the values found by thermal and optical methods agree exactly for the Au acceptor level at 0.15 eV in Ge and for the Au donor levels at 0.20 and 0.30 eV in Ge and Si, respectively (see, for example, [13]). Three aluminum acceptor centers in silicon carbide at 0.28, 0.39, and 0.49 eV were deduced from the photoionization energies and the corresponding thermal ionization energies were 0.27, 0.33, and 0.48 eV [18].

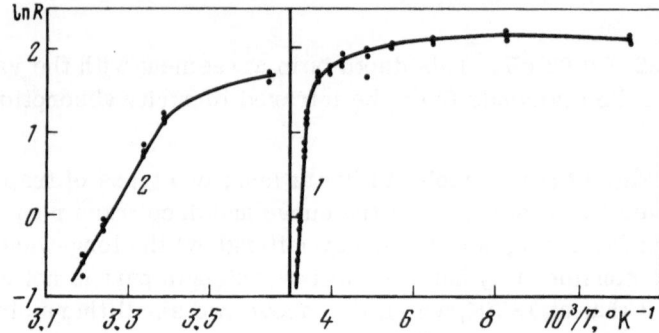

Fig. 15. Dependence of ln R on 1/T for diamond sample A4: 1) complete dependence; 2) high-temperature part of curve 1 on magnified scale.

In the case of shallow traps in Ge and Si there was a small difference (amounting to several percent of the depth) between the optical and thermal ionization energies of these traps [13].

Thus, we may assume that the difference between the optical and thermal depths of traps is manifested only in crystals containing ions. As the ionicity of the bonds decreases, the difference becomes less. This was demonstrated experimentally by R. H. Bube (as reported in [9]) in a study of CdS—CdSe solutions: The difference between the optical and thermal energies for the ionization of a hole from a sensitization center decreased from 0.3 eV for CdS to less than 0.1 eV for CdSe. This reduction in the difference was in agreement with the degree of covalent bonding which was greater for CdSe than CdS.

The value of the difference is equal to the energy of two polarons or one polaron depending on whether the optical liberation process is adiabatic or nonadiabatic.

§ 9. Polarons in ZnS

We shall now consider the nature of polarons in ZnS if their binding energy is ~0.3 eV. There are two theories corresponding to two extreme cases: the theory of large-radius polarons and the theory of small-radius polarons.

The thermal ionization energy of a large-radius polaron is given by the formula (see, for example, [20])

$$E_{th} = -0.054 \frac{m^* e^4 c^2}{\hbar^2} = -1.47 \left(\frac{m^*}{m} \right) c^2 \text{ (eV)}, \tag{22}$$

where $c = (1/\varepsilon_\infty) - (1/\varepsilon)$. Assuming that the relevant parameters of ZnS are $\varepsilon_\infty = 5.07$, $\varepsilon = 8.3$, and $m^*/m = 0.27$, we find that $E_{th} = -2.32 \cdot 10^{-3}$ eV. This value differs very strongly from the experimental results. Therefore, we have to apply the theory of small-radius polarons. This theory was developed by considering the interaction of an electron with optical phonons (see, for example, [21-23]). The polaron binding energy can be calculated if we know the phonon spectrum of a crystal. However, we can estimate approximately this energy using the simplified Mott formula (see, for example, [24]). The polaron energy consists of two parts, one of which is the potential energy of the interaction between ions (E_1) and the other is the potential of an electron in a well (V_p) formed by these ions. We shall simplify the calculations by assuming that the kinetic energy is zero.

Austin and Mott [24] used a simplified model of the potential well of a polaron

$$\left. \begin{array}{l} V_p = -\dfrac{ce^2}{r} \quad \text{for} \quad r > r_p, \\[2mm] V_p = -\dfrac{ce^2}{r_p} \quad \text{for} \quad r < r_p, \end{array} \right\} \tag{23}$$

where r_p is the "polaron radius," and they obtained the following expression for the polaron binding energy

$$E_p = E_1 + V_p = -\frac{1}{2} \frac{e^2 c}{r_p}. \tag{24}$$

If the phonon dispersion can be ignored, the radius r_p is given by the formula

$$r_p = \frac{1}{2} \left(\frac{\pi}{6} \right)^{1/3} \bar{a}, \tag{25}$$

where $(\bar{a})^3$ is the volume per one lattice atom. For example, in the case of a zinc-blende crystal of ZnS, which has four centers per unit cell, we have

$$(\bar{a})^3 = \frac{1}{4}\, a^3,$$

where a is the lattice constant ($a = 5.4$ Å). We find that $\bar{a} = 3.4$ Å and $r_p = 1.37$ Å. The polaron binding energy is then

$$E_p = \frac{(4.8 \cdot 10^{-10})^2 \cdot 6.24 \cdot 10^{11}}{2 \cdot 1.37 \cdot 10^{-3} \cdot 13} = -0.4 \text{ eV},$$

where E_p is the binding energy for an electron at rest. This value is quite close to the measured energy (0.3 eV).

Thus, we may assume that the radii of polarons in ZnS are small. According to the Pekar theory [20], free polarons should absorb infrared radiation energy. In our case this corresponds to $h\nu = 0.45$ eV. This absorption should be correlated with the conductivity of the investigated crystal. The absorption band of ZnS single crystals near 0.5 eV is reported in [25] but it is attributed to shallow levels associated with intrinsic lattice defects.

The authors are grateful to E. E. Bukke, V. V. Antonov-Romanovskii, and E. Yu. L'vova for valuable advice and comments during all stages of the reported investigation.

Literature Cited

1. M. V. Fok, Fiz. Tverd. Tela, 5:1489 (1963).
2. L. A. Vinokurov and M. V. Fok, Opt. Spektrosk., 10:374 (1961).
3. M. V. Fok, Fiz. Tekh. Poluprovodn., 4:1009 (1970).
4. V. V. Antonov-Romanovskii, Kinetics of the Photoluminescence of Crystal Phosphors [in Russian], Nauka, Moscow (1966).
5. V. V. Antonov-Romanovskii and L. A. Vinokurov, Zh. Eksp. Theor. Fiz., 29:830 (1955).
6. V. V. Antonov-Romanovskii and L. A. Vinokurov, Opt. Spektrosk., 1:71 (1956).
7. M. V. Fok, Introduction to the Kinetics of the Luminescence of Crystal Phosphors [in Russian], Nauka, Moscow (1964).
8. F. F. Morehead, J. Phys. Chem. Solids, 24:37 (1963).
9. K. Rebane and V. I. Ruttas, Zh. Prikl. Spektrosk., 6:637 (1967).
10. T. S. Reshetina and V. F. Tunitskaya, Zh. Prikl. Spektrosk., 12:295 (1970).
11. V. V. Antonov-Romanovskii, Dokl. Akad. Nauk SSSR, 20:361 (1938).
12. V. V. Antonov-Romanovskii (Antonov-Romanovsky), Phys. Z. Sowjetunion, 7:366 (1935).
13. R. A. Smith, Semiconductors, Cambridge University Press (1959).
14. A. F. Ioffe, Semiconductors in Modern Physics [in Russian], Izd. Akad. Nauk SSSR, Moscow (1954), p. 322.
15. T. S. Moss, Optical Properties of Semi-Conductors, Butterworths, London (1959).
16. J. A. Elmgren and D. E. Hudson, Phys. Rev., 128:1044 (1962).
17. E. C. Lightowlers, A. T. Collins, P. Denham, and P. S. Walsh, Ind. Diamond Rev., 28:3 (1968)
18. E. E. Bukke, L. A. Vinokurov, I. S. Gorban', A. F. Gumenyuk, Yu. M. Suleimanov, and M. V. Fok, Fiz. Tekh. Poluprovodn., 2:193 (1968).
19. M. Aven and J. S. Prener, Physics and Chemistry of II-VI Compounds, North-Holland, Amsterdam (1967).
20. S. I. Pekar, Investigations in Electron Theory of Crystals [in Russian], Gostekhizdat, Moscow (1951).
21. T. Halstein, Ann. Phys. (Leipz.), 8:343 (1949).
22. M. I. Klinger, Phys. Status Solidi, 11:499 (1965).
23. J. Appel, Solid State Phys., 21:193 (1968).
24. I. G. Austin and N. F. Mott, Adv. Phys., 18:41 (1969).
25. T. S. Reshetina, Opt. Spektrosk., 24:1016 (1968).

NONISOSTRUCTURAL PARAMAGNETIC CENTERS IN ONE-ACTIVATOR CRYSTAL PHOSPHORS

G. E. Arkhangel'skii

The ESR method was used in a study of crystal phosphors activated with one element forming centers of different kinds. It was found that the trigonal Mn^{2+} centers in ZnS:Mn were characterized by $|b_2^0| = 117$ Oe and appeared exclusively in the surface layer of a crystal after the diffusion of the activator at $T \geq 1100°C$. The Eu^{2+} ions in ZnS:Eu formed trigonal and cubic centers and were responsible for characteristic luminescence bands. Strong optical and γ-ray excitation of ruby caused exchange of electrons between the Cr^{3+} centers which resulted in the formation of the Cr^{2+} and Cr^{4+} centers. Crystals of Na_2ZnGeO_4: Mn exhibited two types of Mn^{2+} center, each of which occupied four equivalent positions differing in respect of the orientation of the axes. In this case, the symmetry plane was (100).

One of the main tasks in studies of crystal phosphors is the determination of the individual physical properties of the luminescence centers which largely govern the optical, electrical, and structural properties of crystal phosphors. These characteristics are primarily the charge state of the activator occurring in such centers, symmetry of the surrounding lattice ions, and positions (relative to the allowed bands) of the ground and excited states. The charge state and the symmetry of the environment can be studied by the electron spin resonance (ESR) method which can be used if an impurity ion has an uncompensated net spin.

Investigators prefer to deal with activated crystals in which an impurity ion forms just one type of luminescence center which manifests its properties clearly. However, the synthesis of the majority of real crystals produces, for one reason or another, several nonisostructural centers with different characteristics even when the activating impurity is just one element. In this case, the ESR measurements can give important information on the nature of the centers because the ESR spectra usually overlap less than the luminescence spectra of the same centers. Therefore, it is easier to study the individual centers separately by the ESR method.

The nature of the ESR spectrum, positions of the absorption lines, and their dependences on the orientation of a crystal in a magnetic field can be calculated very accurately using the spin Hamiltonian. This Hamiltonian contains several phenomenological constants whose magnitudes can be found experimentally, for example, from the line positions for a given orientation of a crystal. The substitution of these constants into the Hamiltonian allows us to determine the structure of the ground energy state of a given paramagnetic center and to determine the corresponding wave functions; moreover, we can obtain information on the charge state of the impurity ions, positions of the atoms in the crystal lattice, solubility of the impurity, etc. In some cases, it is sufficient to know these constants to determine the whole structure of the Stark sublevels of a center formed in the electric crystal field. Transition metal ions can be used as effective "probes" in studies of the structure of crystalline solids. The ESR spectra of these ions introduced as small admixtures can be used to detect polymorphic transitions and to estimate the degree of order of the lattice atoms in a crystal.

An important circumstance, which can sometimes be used to attribute with certainty the observed ESR spectrum to a given type of center, is the presence of the nuclear spin of the ion which occurs in this center. The magnetic interaction between the nuclear and electron spins alters slightly the energy of an ion in the magnetic field and this is manifested by the splitting of the energy levels into several hyperfine components. The number of these components, governed by the nuclear spin, is a characteristic "passport" of the paramagnetic ion, which makes it possible to identify centers even in the case of complex ESR spectra. The splitting constant (energy interval) reflects not only the purely magnetic properties of the investigated ion but also the nature of the binding (ionic or covalent) of the impurity ion to the host lattice. This makes it possible to distinguish different paramagnetic centers containing the same impurity ion.

We shall consider three large groups of crystal phosphors activated by one of the transition elements and exhibiting clearly the presence of centers of several types: We shall discuss zinc sulfide doped with manganese and europium, corundum (ruby) doped with chromium ions in different valence states, and sodium zincogermanate doped with manganese. These substances are of interest because they are used as highly efficient phosphors and, at the same time, the activator ions in these substances exhibit relatively easily observed ESR spectra.

CHAPTER I

ESR OF Mn IONS IN ZnS

§ 1. Derivation of Basic Relationships

Electron spin resonance occurs in crystals containing transition group elements because ions of these elements retain a finite magnetic moment of the electron shells even after incorporation into the host lattice. In the majority of the investigated substances, this magnetic moment is of pure spin origin because the electric interactions between the atoms in the condensed phase distort the orbital motion of electrons to such an extent that the orbital magnetic moment is "frozen." When an atom (or ion) has a spin S and practically zero orbital moment, the ground state in an external magnetic field H splits into $2S + 1$ equidistant sublevels with different values of the quantum number M. An alternating electromagnetic field of frequency ν may induce transitions between neighboring levels ($\Delta M = 1$) if H and ν are related by

$$h\nu = g\beta H,$$

where β is the Bohr magneton and $g = 2.0023$. Transitions from the upper sublevel to the lower occur with the same probability as from lower level to the upper. However, if the investigated substance is initially in thermal equilibrium, the population of the lower level is higher and the sample as a whole absorbs energy. Relaxation phenomena may establish a steady state in which the energy of the alternating field is converted into heat. The equidistant distribution of the energy levels should give rise to just one absorption line.

The ESR spectra of real crystals depend strongly on small changes in the magnetic moment of a particle due to the orbital motion of electrons. Therefore, in ordered atomic systems (crystals), the resonance conditions depend on the orientations of the bonds and crystal axes in a magnetic field and, moreover, the spin level is split even in the absence of an external magnetic field. The magnetic sublevels become nonequidistant and several absorption lines are observed. However, the theoretical approach to such a paramagnetic system remains purely of the "spin" type and the transition energies can be calculated from the effective spin which usually is equal to the true spin. The spin Hamiltonian used in such calculations is a polynomial which depends only on the projections S_x, S_y, and S_z of the spin onto the magnetic field direction and on certain constant parameters, which include the external magnetic field. This Hamiltonian does not contain explicit orbital variables, which simplifies considerably the solution of the corresponding Schrödinger equation for finding the spin sublevels. In general, the spin Hamiltonian depends primarily on the following parameters: the external magnetic field H, electron S and nuclear I spins of the atom, constant A representing interaction between the electron shells and the nuclear spin, parameters b_m^n representing the initial splitting of the levels in the crystal field, g factor which may differ from 2.0023 for an ion in the crystal lattice, and angles θ between the magnetic field direction and the crystal field axes. The same parameters, together with the magnetic quantum numbers M and m (representing electron and nuclear subsystems), govern the positions of the spin sublevels, which are the eigenvalues of the Hamiltonian.

Having found these sublevels, we can calculate the ESR spectrum allowing for the fact that, according to the selection rules, the value of M should change by unity and m should remain unaffected in the investigated transition regions; conversely, we can calculate the parameters g, b_m^n, and A of centers of a given type from the experimentally obtained ESR spectra. The spins S and I can be deduced from the number of lines in the spectrum and other parameters from the resonance fields.

In a crystal field of axial (trigonal) symmetry, characterized by the parameter b_2^0, if the value of g is independent of the magnetic field direction and the field is so strong that

$$|b_2^0| \ll H, \quad |A| \ll H, \tag{1.1}$$

the system of spin levels $E_{M,m}$ can be described quite accurately by the following relationships [1, 2], which include isotropic (E_i) and anisotropic (E_a) parts:

$$E_{M,m} = E_i + E_a. \tag{1.2}$$

Here,

$$E_i = g\beta H M + g\beta A M m + \frac{(g\beta A)^2}{2h\nu} \{M[I(I+1) - m^2] - m[S(S+1) - M^2]\},$$

$$E_a = \frac{1}{6} b_2^0 C_0' (3\cos^2\theta - 1) + \frac{1}{480} b_4^0 C_0'' (35\cos^4\theta - 30\cos^2\theta + 3) -$$
$$- r'C_1 \sin^2\theta \cos^2\theta + \frac{1}{4} r'C_2 \sin^4\theta + r'C_3 m \frac{g\beta A}{h\nu} \sin^2\theta \cos^2\theta + \frac{1}{4} r'C_4 m \frac{g\beta A}{h\nu} \sin^4\theta,$$

where $H_0 = h\nu/g\beta$ and $r' = (g\beta b_2^0)^2/2h\nu$ (the constants b_2^0, b_4^0, and A are in units of the magnetic field):

$$
\left.
\begin{aligned}
C_0' &= 3M^2 - S(S+1); \\
C_0'' &= 35M^4 - 30S(S+1)M^2 + 25M^2 - 6S(S+1) + 3S^2(S+1)^2; \\
C_1 &= 8M^3 + M - 4MS(S+1); \\
C_2 &= 2M^3 + M - 2MS(S+1); \\
C_3 &= \frac{1}{M}\{[M^2 - S(S+1)]^2 - M^2\}; \\
C_4 &= M[(2M^2 - 1) - 2S(S+1)].
\end{aligned}
\right\}
\tag{1.3}
$$

The magnetic fields corresponding to the allowed transitions from states with $M-1$ and m to states with M and m have the values

$$H_{M,m} = H_0 - Am - \frac{A^2}{2H_0}[I(I+1) - m^2 + m(2M-1)] -$$
$$- \frac{1}{2}b_2^0(2M-1)(3\cos^2\theta - 1) - \frac{1}{480}b_4^0 a_0(35\cos^4\theta -$$
$$- 30\cos^2\theta + 3) + ra_1\sin^2\theta\cos^2\theta - \frac{1}{4}ra_2\sin^4\theta -$$
$$- r\frac{A}{H_0}a_3 m\sin^2\theta\cos^2\theta - \frac{1}{4}r\frac{A}{H_0}a_4 m\sin^4\theta, \tag{1.4}$$

where

$$r = (b_2^0)^2/2H_0, \quad a_0 = C''_{0_M} - C''_{0_{M-1}}, \quad a_1 = C_{1_M} - C_{1_{M-1}},$$
$$a_2 = C_{2_M} - C_{2_{M-1}}, \quad a_3 = C_{3_M} - C_{3_{M-1}}, \quad a_4 = C_{4_M} - C_{4_{M-1}}.$$

The magnetic fields corresponding to the forbidden transitions from states with $M-1$ and m to states with M and $m-1$ have the values

$$H_{M,m-1} = H_0 - A(-M+m) - \frac{A^2}{2H_0}[4Mm - m - m^2 + I(I+1) + S(S+1) -$$
$$- M(M+1)] - b_2^0\left(M - \frac{1}{2}\right)(3\cos^2\theta - 1) - \frac{b_4^0}{480}a_0(35\cos^4\theta -$$
$$- 30\cos^2\theta + 3) + ra_1\sin^2\theta\cos^2\theta - \frac{r}{4}a_2\sin^4\theta - \frac{rA}{H_0}(ma_3 -$$
$$- C_{3_M})\sin^2\theta\cos^2\theta - \frac{rA}{4H_0}(ma_4 - C_{4_M})\sin^4\theta. \tag{1.5}$$

The magnetic fields corresponding to the forbidden transitions from states with $M-1$ and $m-1$ to states with M and m have the values

$$H_{M,m} = H_0 - A(M+m-1) - \frac{A^2}{2H_0}[-m^2 + m + I(I+1) - S(S+1) +$$
$$+ M(M-1)] - b_2^0\left(M - \frac{1}{2}\right)(3\cos^2\theta - 1) - \frac{b_4^0}{480}a_0(35\cos^4\theta -$$
$$- 30\cos^2\theta + 3) + ra_1\sin^2\theta\cos^2\theta - \frac{r}{4}a_2\sin^4\theta - \frac{rA}{H_0}(ma_3 +$$
$$+ C_{3_{M-1}})\sin^2\theta\cos^2\theta - \frac{rA}{4H_0}(ma_4 + C_{4_{M-1}})\sin^4\theta. \tag{1.6}$$

The anisotropic component of the resonance fields H_a can generally be represented in the simpler form

$$H_a = - b_2^0\left(M - \frac{1}{2}\right)(3\cos^2\theta - 1) - \frac{b_4^0 a_0}{480}(35\cos^4\theta - 30\cos^2\theta + 3) - r(p\sin^2\theta - q\sin^4\theta), \tag{1.7}$$

where the parameters p and q are different for the allowed and forbidden transitions:

$$p_1 = a_1 - \frac{A}{H_0}a_3 m, \quad q_1 = a_1 + \frac{a_2}{4} + \frac{A}{H_0}\left(\frac{a_4}{4} - a_3\right)m \tag{1.8}$$

for the allowed transitions from states with $M-1$ and m to states with M and m;

$$p_2 = a_1 - \frac{A}{H_0}(ma_3 + C_{3_{M-1}}), \quad q_2 = a_1 + \frac{a_2}{4} + \frac{A}{H_0}\left[\frac{1}{4}(ma_4 + C_{4_{M-1}}) - ma_3 - C_{3_{M-1}}\right] \tag{1.9}$$

for the forbidden transitions from states with $M-1$ and $m-1$ to states with M and m;

$$p_3 = a_1 - \frac{A}{H_0}(ma_3 - C_{3_M}),$$
$$q_3 = a_1 + \frac{a_2}{4} + \frac{A}{H_0}\left[\frac{1}{4}(ma_4 - C_{4_M}) - ma_3 + C_{3_M}\right] \tag{1.10}$$

for the forbidden transitions from states with M − 1 and m to states with M and m − 1.

The characteristics of the ESR spectrum of a powder sample, associated with the anisotropy with the original spectrum of the crystal, can be found by averaging the value of H over the angles θ within the range $0 \le \theta \le \pi/2$.

We shall consider a fine-structure line due to the $M = -\frac{1}{2} \to \frac{1}{2}$ transition, which is the strongest component of the hyperfine structure corresponding to the lowest fields. The anisotropy of this component is governed entirely by the terms containing r in Eq. (1.4). If the quantity $(b_2^0)^2/2H_0$ exceeds considerably the line width for a single particle (microcrystal) in a powder, a random distribution of particles (in a polycrystalline sample) should give rise to a broadened line of quite different profile. The profile of this line can be found by estimating the relative intensity of the absorption in each elementary interval of fields H near the line center. This problem simplifies if the original ESR line of a single particle is approximated by a Π-like line of width equal to the half-width of the original line. In this case, the absorption intensity is proportional to the number of particles making the same contribution to the total absorption in such an elementary field interval. Disordered particles N_0, whose crystal field axes are oriented within the angle $\delta\theta$, are distributed between the angles θ relative to a selected direction (in the present case, the direction of the external magnetic field) in accordance with the expression

$$dN = N_0 \sin\theta\, d\theta. \tag{1.11}$$

If this expression is modified by replacing θ with H in accordance with Eq. (1.4) and the result is then integrated with respect to dH within the limits of the line width ΔH, the required number of particles considered as a function of the magnetic field is given by

$$N = N_0 \left[1 - \frac{p_1}{2q_1} - \left(\frac{p_1^2}{4q_1} - \frac{H}{rq_1} \right)^{1/2} \right]^{1/2} \Big|_H^{H+\Delta H} \tag{1.12}$$

in the field interval $(p_1 - q_1)\,r \le H < H_{\parallel}$ and by

$$N = N_0 \left[1 - \frac{p_1}{2q_1} - \left(\frac{p_1^2}{4q_1} - \frac{H}{rq_1} \right)^{1/2} \right]^{1/2} \Big|_H^{H+\Delta H} + N_0 \left[1 - \frac{p_1}{2q_1} + \left(\frac{p_1^2}{4q_1} - \frac{H}{rq_1} \right)^{1/2} \right]^{1/2} \Big|_H^{H+\Delta H} \tag{1.13}$$

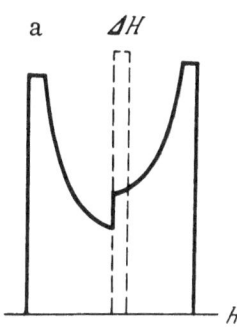

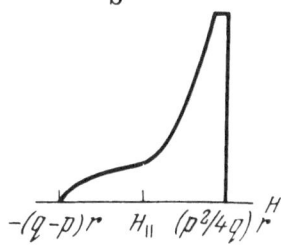

Fig. 1. Calculated profile of an anisotropically broadened Π-like line of the allowed $M = -\frac{1}{2}$, $m \to M = \frac{1}{2}$, m (a) and forbidden $M = -\frac{1}{2}$, $m \to M = \frac{1}{2}$, m − 1 (b) transitions.

in the field interval $H_\parallel \le H \le (p_1^2 r / 4q_1) - \Delta H$. It follows from the above dependence (shown graphically in Fig. 1a) that the average component of the hyperfine structure is broader, because of the anisotropy, than the original line, and it gives rise to two pronounced maxima [3] separated by

$$\delta H = \frac{r\left(\frac{p_1}{2} - q_1\right)^2}{q_1}. \qquad (1.14)$$

The separation δH is equal to the difference between the extremal resonance fields corresponding to one particle. The absorption intensity is approximately the same at the two maxima, and it follows from Eqs. (1.12) and (1.13) that this absorption is given by

$$I_1 \approx \frac{I_{0_1}}{b_2^0} \sqrt{\frac{\Delta H H_0}{q_1 - \frac{p_1}{2}}} \qquad (1.15)$$

[if $(b_2^0)^2 \gg 0.3 H_0 \Delta H$], where I_{0_1} is the total intensity of the original unbroadened line. The expressions (1.14) and (1.15) are valid for all the hyperfine-structure components of the $M = -1/2 \to +1/2$ transition, but the values of δH and I_1 are different, since the parameters p_1 and q_1 include the quantum number m.

The fine-structure components of the $M = 1/2 \to 3/2$ transition in a powder sample are broadened more strongly because resonance fields of the particles are distributed within the range $H \approx 3b_2^0$. In this case, each of the components has only one maximum at the field

$$H = H_0 + b_2^0 - \frac{1}{160} b_4^0 a_0 + r(p_1 - q_1) - Am - \frac{A^2}{2H_0}[I(I+1) - m^2 + 2m], \qquad (1.16)$$

which corresponds to the perpendicular orientation of the c axis of one microcrystal relative to the magnetic field. The line profile is described by

$$I = I_0\left[\sqrt{\frac{\Delta H}{b_2^0} - \frac{H}{b_2^0} + 1} - \sqrt{1 - \frac{H}{b_2^0}}\right]. \qquad (1.17)$$

The absorption intensity at the maximum is considerably less than for the $M = -1/2 \to 1/2$ line and is given by

$$I_2 = I_{0_2} \sqrt{\Delta H / 3b_2^0}, \qquad (1.18)$$

where I_{0_2} is the total intensity of the corresponding unbroadened line or

$$I_2 = I_1 \frac{(2S+1)^2 - 4}{(2S+1)^2} \sqrt{\frac{b_2^0\left(q_1 - \frac{p_1}{2}\right)}{3H_0}}. \qquad (1.19)$$

The forbidden transitions $M = -1/2, m \to 1/2, m-1$ and $M = -1/2, m-1 \to 1/2, m$ are responsible (if the forbiddenness is lifted) to 4m additional lines in the ESR spectrum, and the positions of these lines are described by Eqs. (1.5) and (1.6). The intensities of these lines depend on the orientation of the crystal field axes relative to the external magnetic field and on the value of the constant b_2^0 (see [3]):

$$I \propto \left(\frac{b_2^0}{H} \sin 2\theta\right)^2. \qquad (1.20)$$

TABLE 1

Type of center	g	$\lvert b_2^0 \rvert$, Oe	Sign of b_2^0	A, Oe	Ref.
Cubic	2.0025 ± 0.0005	0		-68.6 ± 0.3	[4]
Trigonal:					
I	2.0025 ± 0.0005	142.4 ± 0.3	—	-68.6 ± 0.3	[4, 5]
II	2.0025 ± 0.0005	40.3 ± 0.9	Not det.	-68 ± 0.3	[5, 6]
III	2.0016 ± 0.0001	113.2 ± 2	—	-70 ± 1	[7]

Therefore, in contrast to the allowed transition M = $-^1/_2$, m $\rightarrow$ $^1/_2$, m, which has two maxima, each of the forbidden transition lines of a powder sample should have only one maximum (Fig. 1b) at the field

$$H = H_0 - A(-M+m) - \frac{A^2}{2H_0}\left[m^2(4M-1) + I(I+1) + S(S+1) - M(M+1)\right] + \frac{p_3^2}{4q_3} \qquad (1.21)$$

and at

$$H = H_0 - A(M+m-1) - \frac{A^2}{2H_0}[-m^2 + m + I(I+1) - S(S+1) + M(M+1)] + \frac{p_2^2}{4q_2}, \qquad (1.22)$$

where p_2, p_3, q_2, and q_3 are the parameters given by Eqs. (1.9) and (1.10).

The Mn^{2+} ion has the ground state $^6S_{5/2}$. In a magnetic field, the sixfold spin degeneracy of this state is fully lifted and, since the resultant sublevels are generally nonequidistant (because of the additional splitting by the crystal field), the ESR spectrum of a single crystal containing Mn^{2+} ions usually consists of five fine-structure lines due to the $\Delta M = 1$ transitions. Each of these lines consists of six hyperfine components corresponding to different values of m (I = 5/2).

A comparison of the parameters representing the interaction of manganese centers with the ZnS lattice field, reported in several papers, shows that cubic centers of one type and trigonal centers of several distinct types are found in crystals grown by different methods and under different conditions. The principal parameters of these centers are listed in Table 1.

All the possible trigonal centers have not yet been observed in one crystal, but it is usual to find trigonal as well as cubic centers, which is evidence of the inhomogeneity of the structure of "hexagonal" crystals. The differences between the values of b_2^0 of the various trigonal centers are probably due to the differences in the lattice structure near these centers, for example, due to variations in the sequence of layers. Therefore, in the identification of the manganese centers, it is desirable to find the value of b_2^0 using hexagonal crystals with the fewest defects. Such crystals are, for example, zinc sulfide powder particles prepared at a temperature slightly higher than the sphalerite−wurtzite transition point. These were the particles used in our determination of the parameter b_2^0 of the Mn^{2+} ion in ZnS.

§2. Interpretation of ESR Spectra and Determination of b_2^0 of Hexagonal ZnS:Mn

The ESR spectra of ZnS:Mn and of all the other samples investigated in the present study were recorded at ν = 9250 MHz. We used an rf spectrometer, which plotted either the absorption curve or its first derivative. The first derivative was obtained employing the usual direct-amplification scheme with high-frequency modulation of the external magnetic field. The mod-

ulation frequency was 120 kHz, and the modulation amplitude was within the range 0–3 Oe.
A tunable cylindrical reflection-type resonator formed an arm of a microwave bridge circuit
and was excited by the H_{011} mode. The modulation coil was placed inside the resonator. A
detected ESR signal was amplified by a narrow-band amplifier ($k = 10^5$, $\Delta f = 2$ kHz) and recti-
fied with a phase-sensitive rectifier. The frequency of a klystron oscillator was tuned auto-
matically to the resonator frequency.

The absorption spectrum was recorded using a spectrometer variant with a low-frequency
(50 Hz) modulation of the magnetic field to a depth exceeding the line width at its base; in this
case, oscillographic recording was employed. Noise was reduced by additional high-frequency
amplitude modulation of the ESR signal with the aid of a controlled crystal diode and amplification with a band-pass amplifier [8].

The low-temperature ESR spectra were recorded using a quartz Dewar flask with an
unsilvered branch tube of 12 mm diameter at the base (this tube contained a sample and was
inserted into the resonator) or a glass cryostat placed entirely inside the resonator.

The resonance magnetic field (250–9800 Oe) was measured with a fluxmeter utilizing a
probe which detected the NMR of protons and Li^7 nuclei; this probe was placed near the resonator containing the sample.

The above analysis of the profile of anisotropically broadened lines and estimates of
their relative intensities made it possible to identify the origin of the principal lines in the
ESR spectrum of ZnS:Mn powder with the hexagonal lattice structure. This spectrum consisted of a large number of lines of different intensities (Fig. 2). It manifested the sixfold
repetition of lines of approximately the same intensity, which were characteristic of the hyperfine structure (lines in groups I, II, and IV). In addition to these sixfold-repeated lines, the
spectrum included 10 additional lines grouped in pairs (lines of group III). Group I lines were
attributed to the hyperfine structure of the $M = -^1/_2 \rightarrow {}^1/_2$ line, each of whose components was
split into two parts because of the anisotropy, whereas weak lines belonging to group II were
attributed to the hyperfine structure of the symmetrically located $M = {}^1/_2 \rightarrow {}^3/_2$ and $M = -^3/_2 \rightarrow$
$-^1/_2$ lines. The positions of these line groups in the spectrum made it possible to determine
the constant b_2^0 by two independent methods using Eqs. (1.14) and (1.16). The results of this
determination were compared (Table 2) with the values of b_2^0 deduced from the ESR spectrum
of a ZnS:Mn single crystal. The hyperfine-structure constant A, occurring in the original
formula (1.14) via the parameters p_1 and q_1, was calculated from

$$A = \frac{1}{5}(H_{5/2} - H_{-5/2}),\tag{1.23}$$

where the values of $H_{5/2}$ and $H_{-5/2}$ were the average values of the H for the broadened (and
split into two) components m = 5/2 and m = –5/2 of group I (Fig. 2).

Possible errors in Table 2 were estimated by comparing the values of the constant b_2^0
for Mn and $CaCO_3$ (hexagonal lattice) obtained for a powder and a single crystal [3]. The best

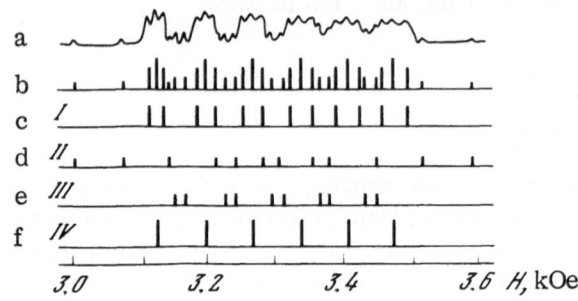

Fig. 2. ESR spectrum of a ZnS:Mn powder
with hexagonal structure: a) absorption
spectrum; b) schematic representation of
the spectrum; c) lines of the $M = -^1/_2 \rightarrow$
$+^1/_2$ transition; d) lines of the $M = -^3/_2 \rightarrow$
$-^1/_2$ and $M = {}^2/_1 \rightarrow {}^3/_2$ transition; e) lines
of the forbidden transitions with $\Delta m = 1$;
f) lines due to cubic centers.

TABLE 2

Sample	M	m	I/I_0	A, Oe	$\|b_2^0\|$, Oe	b_2^0, Oe (single crystal)
ZnS-Mn	$1/2$	$-3/2$	0.23	-68.5	118	-140.1 [4—6]
Hexagonal	$3/2$	$-5/2$	0.04	-68.5	117	
CaCO$_3$-Mn	$1/2$	$-5/2$		-93.5	90	85.5 [3]

agreement between the calculated spectrum of the hexagonal ZnS:Mn with the experimental results was obtained for $|b_2^0| = 117.5$ Oe and $A = -68.5$ Oe. A deviation of b_2^0 by ± 2 Oe shifted the calculated lines by an amount considerably greater than the widths of the observed lines. The measured ratios of the intensities of the lines in the groups mentioned above were in good agreement with the calculated results, which confirmed that these groups were due to the $M = -1/2 \rightarrow 1/2$, $M = 1/2 \rightarrow 3/2$, and $M = -3/2 \rightarrow -1/2$ transitions.

The lines of group III in the ESR spectrum (Fig. 2) could be described quite satisfactorily by Eqs. (1.21) and (1.22). The calculated resonance fields for the two low-field lines in this group differed from the experimental values by 2-3 Oe. Therefore, the lines in group III were attributed to the forbidden transitions $M = -1/2$, $m \rightarrow 1/2$, $m - 1$ and $M = -1/2$, $m - 1 \rightarrow 1/2$, m in the Mn^{2+} ion.

In addition to the lines mentioned above, the spectrum of the hexagonal ZnS:Mn powder included also several values which could not be explained by the anisotropy or forbidden transitions. These lines formed group IV in Fig. 2. Their positions in the spectra were described satisfactorily by

$$H = H_0 - Am - \frac{A^2}{2H_0}\left(\frac{35}{4} - m^2\right),$$ (1.24)

where $g = 2.0018$ and $A = -68.5$ Oe. These constants corresponded to the ESR spectrum of Mn^{2+} in the cubic ZnS, well known from studies of powders and single crystals [4, 9]. In this case, we found that $b_2^0 = 0$, and the lines were not split into two components by the anisotropy; moreover, the forbidden transition lines were not observed. The presence of the lines of group IV indicated that zinc sulfide powder heated at 1150°C transformed only partly to the hexagonal modification and contained some cubic phase. A comparison of the intensities of the group I and group II lines belonging exclusively to the hexagonal modification, carried out subject to Eq. (1.19), and a comparison of the intensities of these lines with the spectrum of the cubic ZnS:Mn, applied to samples with the same amounts of manganese, demonstrated that the proportion of the cubic phase in the investigated hexagonal ZnS:Mn powders did not exceed 15%.

It follows from the above interpretation of the ESR lines of ZnS:Mn that: 1) the manganese impurity in purely hexagonal ZnS gives rise to paramagnetic centers of just one type; 2) the constant b_2^0 of these centers, representing the axial component of the crystal field, differs considerably from the constant of the trigonal (type I) Mn^{2+} centers (Table 1) in predominantly cubic single crystals; 3) the highest concentration of the hexagonal phase in ZnS powders is 85-90%; 4) the Mn^{2+} centers in the hexagonal powder are the trigonal type III centers (Table 1).

The difference between the values of b_2^0 of the trigonal centers in the cubic and hexagonal crystals is clearly due to a considerable distortion of the crystal field in polycrystalline ZnS by the ions of accidental impurities diffusing into the crystal from the atmosphere during heating. These are most likely oxygen ions, reported in [10], after heating of zinc sulfide powder under conditions similar to those employed in our study. Although in both cases the samples were heated in a strongly reducing atmosphere, the lattice could still absorb large amounts of oxygen because the total surface of the finely powdered material was large.

The crystal field corresponding to a given type of center causes a corresponding Stark splitting of the higher orbital levels ^4G, ^4P of the Mn^{2+} ion, which govern the optical properties of these centers (absorption spectrum and visible luminescence). In those cases when a crystal contains different types of center in comparable amounts, the overall spectrum should be a superposition of the elementary bands of the individual centers. The experimentally obtained complex luminescence spectra of manganese-activated ZnS crystals may be explained by the presence of centers of these types.

It should be pointed out that crystals of synthetic wurtzite, investigated in [7], were grown as thin plates, and a considerable proportion of the hexagonal phase in these crystals could be due to the relatively thick surface layer where the defect concentration was highest.

§ 3. Influence of Manganese Impurities on the Formation of Hexagonal ZnS

It follows from the results of several investigations that the stabilization of the various ZnS polytypes is influenced strongly by deliberately introduced impurities, as well as accidental impurities which are easy to identify but difficult to remove. A small amount of oxygen or aluminum present during the heating of a phosphor gives rise to a considerable proportion of the hexagonal phase [11, 12], whereas the introduction of copper and silver ions favors the formation of a predominantly cubic lattice [13]. In some cases, this may be due to the structure of the sulfide in question (or to the formation of zinc oxide in the case of oxygen). This influence can be explained bearing in mind that the free energies of the cubic and hexagonal modifications of zinc sulfide are almost equal. Therefore, even a slight change in the free energies of these modifications, caused by the introduction of an impurity, may affect the stability of a given modification at the selected temperature. This is why second heating usually fails to alter significantly the new crystalline modification unless we use the sublimation method in which the crystal lattice dissociates completely. In the latter case, the presence of, for example, manganese favors the hexagonal modification at lower heating temperatures [14]. The activation of ZnS powder at such temperatures produces a material which is almost entirely hexagonal, whereas a single crystal forms at this temperature with just a small amount of hexagonally packed layers. This dependence of the lattice structure of activated crystal phosphors on the geometric dimensions of the crystals suggests that the formation of the hexagonal phase in ZnS at sufficiently high temperatures is influenced significantly by the processes occurring on the surface.

We determined the influence of these processes by investigating the disturbed surface layer formed in the process of activation of cubic crystals [15]. Preliminary experiments established that the thickness of the hexagonal surface layer was slight even after prolonged activation. Therefore, a reliable estimate of the volume of this layer was obtained by using large numbers of fine crystals with a considerably greater surface area than one single crystal of the same mass. Such fine crystals were prepared by crushing unactivated crystals with a predominantly cubic structure. These crystals were grown from the vapor phase in a closed ampoule at 1380°C in an H_2S or NH_3 atmosphere. The rate of pulling of this ampoule through the hottest zone was 0.18 mm/h. The grown crystals were cooled to room temperature in 120 min. These crystals were pulverized in a porcelain mortar. They were then activated with manganese from $MnCl_2$ at 1170°C applied for 30 min in the presence of 1% of a chloride (NaCl) flux; the activation was performed in a dry ammonia atmosphere. The resultant mixture was separated into three fractions by passing it through sieves. The relative granulometric composition was determined for each fraction by counting the number of particles with different transverse dimensions in a standard sample. This was done by taking a known amount

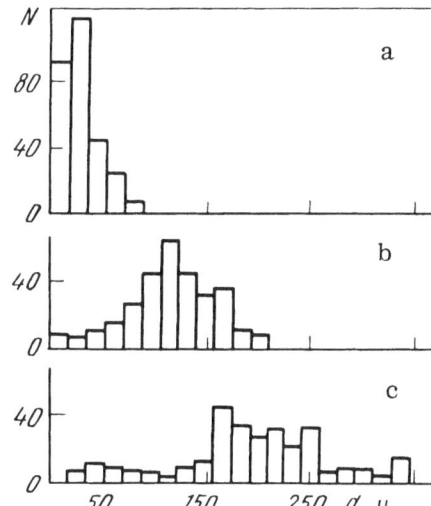

Fig. 3. Relative granulometric composition of the fine (a), intermediate (b), and coarse (c) fractions of ZnS.

(about 5 mg) of one fraction and spreading it uniformly on an area of about 100 mm² on the microscope stage. About 200-250 particles were found to be within the field of view of the microscope, and their dimensions, which ranged from largest to smallest, were then determined. The dimensions of the particles were estimated using a grid with divisions of 0.005 mm. We calculated the number of particles whose "diameters" did not differ by more than 10 μ. The three fractions had the largest numbers of particles with "diameters" of 25, 100, and 210 μ, respectively (Fig. 3).

The number of Mn^{2+} ions present simultaneously in two crystalline phases of the same sample was determined by measuring first the normalized ESR line intensities of the cubic and hexagonal phases. These measurements were carried out using separate samples of cubic and hexagonal structures with known amounts of manganese. The spectral lines, typical of the hexagonal phase, were the lines due to the $\Delta m = \pm 1$ forbidden transitions which did not overlap the cubic-phase lines. The relative amounts of Mn^{2+} ions were calculated on the assumption that the ESR line profiles and widths did not vary greatly with the impurity concentration and the number of impurity ions was proportional to the line intensity. This number was deduced from the formula

$$n_{Mn^{2+}} \propto k U_{hex.forb} + U_{cub},$$ (1.25)

where $U_{hex.forb}$ is the intensity of the forbidden ESR line emitted by the investigated sample, and U_{cub} is the intensity of the cubic-phase line of the same sample (Fig. 4); k is a normalizing coefficient found experimentally by comparing the ESR spectra of cubic and hexagonal samples with the same amounts of manganese (k = 8.9).

A comparison of the ESR spectra of the fractions under consideration shows that the ratio of the intensities of the lines typical of the hexagonal and cubic phases is largely independent

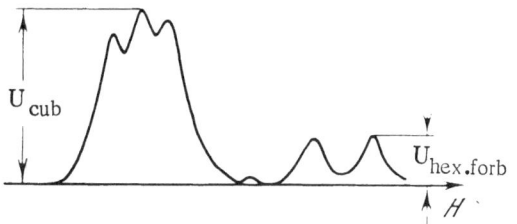

Fig. 4. ESR lines of ZnS:Mn used to determine the concentration of manganese. Here, $U_{hex.forb}$ is the intensity of the forbidden transition line in the hexagonal phase, and U_{cub} is the intensity of the cubic-phase line.

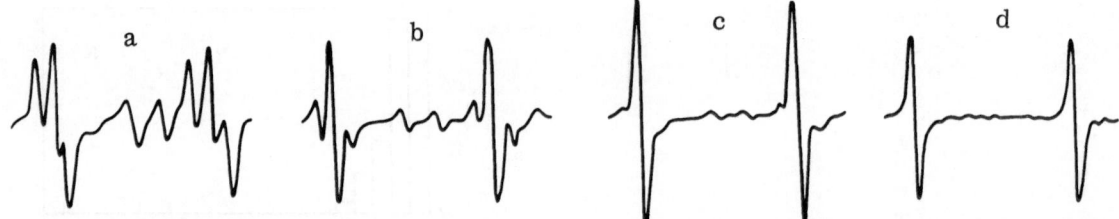

Fig. 5. ESR spectra of ZnS:Mn powder: a) fine fraction; b) intermediate fraction; c) coarse fraction; d) intermediate fraction after partial dissolution.

of the particle size.   This confirms the hypothesis that the formation of the hexagonal phase during the activation of zinc sulfide begins at the surface. This is supported by the observation that a partial dissolution of single crystals reduces the intensity of the ESR spectrum of the hexagonal phase (this intensity falls nearly to zero). Figure 5 shows the low-field parts of the spectra of the fine, intermediate, and coarse fractions, as well as of the intermediate fraction subjected to partial dissolution. It is clear from Fig. 5 that the hexagonal phase does not appear in the ESR spectrum of sample d.

An estimate of the volume (thickness) of the surface layer rich in the hexagonal phase can be obtained by comparing the characteristic components of the spectra, provided the distribution of the manganese ions in each microcrystal is known; this distribution may not be homogeneous because of incomplete diffusion. We found the distribution of manganese by the method of successive dissolution of the sample and repeated measurements of the total ESR line intensity after each dissolution. A certain amount of each fraction of fixed weight was subjected to the action of a solvent (10% solution of HCl at its boiling point). The proportion of the sample remaining after dissolution was determined by weighing. A standard sample (75 mg) was then taken from this residue, and the intensity of the total ESR spectrum was determined relative to a standard DPPH sample. Variation of the dissolution process produced a series of samples for each fraction which were of identical weight and from which surface layers of different thickness were removed by this process.

The measured manganese distributions are plotted in Fig. 6, showing the dependence of the total amount of Mn^{2+} on the mass of a sample P during its dissolution. In the case of a homogeneous distribution, this dependence should be linear. It follows from Fig. 6 that the Mn^{2+} ions in the grains of the intermediate fraction were distributed almost homogeneously, whereas the coarse-fraction grains exhibited a considerable concentration gradient. The distribution of

Fig. 6. Relative amounts of manganese remaining after partial dissolution of ZnS:Mn. Here, P_i and P_d are the weights (masses) of a sample before and after dissolution; the crosses represent the coarse fraction and the circles, the intermediate fraction; the continuous curves are calculated for a homogeneous distribution (1) and for the intermediate and coarse fractions (2 and 3, respectively).

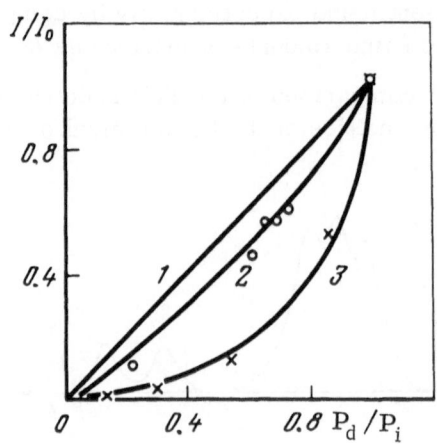

TABLE 3

Fraction	K, μ^3	L, μ^2	M, μ	n
Fine	4 760	218	12.1	280
Intermediate	199 000	378	58.2	300
Coarse	253 000	9 950	92.3	291

manganese in the fine-fraction particles should be even more homogeneous than in the intermediate fraction.

The considerable inhomogeneity of the impurity distributions in the intermediate and coarse fractions enabled us to estimate the diffusion coefficient under the selected heat conditions during activation. This estimate was based on Fick's law extended to the case of impurity diffusion into a spherical sample through its surface [16]. The continuous curves in Fig. 6 represent theoretically calculated amounts of manganese ions in these samples for spheres of decreasing diameter in the course of partial dissolution. The best agreement with the experimental results was obtained for the diffusion coefficient D = 2.2 · 10^{-9} cm^2/sec (at 1200°C). The scatter of the experimental points corresponded to a variation of D by a factor of 1.5-2. The results obtained indicated that the best samples for the investigation of the formation of the hexagonal phase under the selected heating conditions were the fine and intermediate fractions.

The thickness of a hexagonal layer appearing on the surface of a ZnS microcrystal during activation with manganese was estimated by two methods. The first method is based on a comparison of the ratio of the amounts of the cubic and hexagonal phases in the samples of the three fractions considered above. This ratio increases with the particle size because the relative proportion of the volume occupied by the surface layer decreases with increasing size. This ratio differs by several tens between the coarse and fine fractions, as indicated by the ratio of the intensities of the corresponding ESR spectra (Fig. 5). In calculating the effective thickness Δ of a hexagonal layer, each grain is regarded as a sphere of diameter equal to the transverse size of this grain. In general, the value of Δ is related to the radii r_i of spherical particles, differing somewhat within one fraction, by the following expression:

$$\frac{\sum_i (r_i - \Delta)^3 n_i}{\sum_i [r_i^3 - (r_i - \Delta)^3] n_i} = R, \tag{1.26}$$

where n_i is the number of particles of radius r_i in a given fraction; R = V_{in}/V_s is the ratio (deduced from the ESR spectra) of the amounts of Mn on the inner part of a sphere with cubic structure to the amount in the surface layer.* The value of Δ can be found from a relationship which follows from Eq. (1.26):

$$\Delta^3 - 3M\Delta^2 + 3L\Delta - \frac{K}{R+1} = 0, \tag{1.27}$$

where the coefficients

$$M = \frac{1}{n}\sum_i r_i n_i, \qquad L = \frac{1}{n}\sum_i r_i^2 n_i, \qquad K = \frac{1}{n}\sum_i r^3 n_i$$

---

* As mentioned earlier, removal of the surface layers suppressed completely the ESR signal due to the hexagonal phase.

are the average values of the particle radius, its square, and cube; n is the total number of investigated particles in a given fraction. These values can be calculated from the results of a granulometric analysis of each fraction and are given in Table 3.

The second method of determining Δ involves the partial removal of the surface layer by dissolution. As described above, the dissolution method was used to determine the homogeneity of the distribution of the Mn^{2+} ions diffused into the microcrystals. The results of a quantitative determination of the ratio of the masses of the cubic and hexagonal forms of ZnS in the three fractions were also used to calculate Δ. After partial removal of the surface layer δ from a sphere, the value of Δ is related to the radii of the spherical particles by

$$\frac{\sum_i [(r_i - \delta)^3 - (r_i - \Delta)^3] n_i}{\sum_i [r_i^3 - (r_i - \Delta)^3] n_i} = R', \tag{1.28}$$

where $R' = V_d/V_s$ is the ratio (found from the ESR spectra) of the volumes of the hexagonal layer remaining after partial dissolution and of the hexagonal surface layer present initially. In the calculation of Δ, this expression is reduced to the form

$$\Delta^3 - 3M\Delta^2 + 3L\Delta - \frac{1-s}{1-R'} K = 0, \tag{1.29}$$

where K, L, and M are the coefficients defined earlier; $s = P_d/P_i$ is the ratio of the masses of the partly dissolved and original samples. The value of s was found by weighing on the assumption that the densities of the hexagonal and cubic modifications differed only slightly.

The measurements carried out in this way on the three fractions demonstrated that the thickness of the hexagonal layer formed on crystals during the diffusion of manganese was 10 μ, which was practically in agreement with the results obtained by the first method (Table 4).

Allowance for the inhomogeneous distribution of manganese in the intermediate fraction gave a correction of the order of 10%, which was considerably less than the experimental error amounting to about 50% (deduced from a comparison of the results obtained by the two methods described above).

The correction needed for the coarse fraction was 25% (the values should be reduced by this amount). The second method was found to be unsuitable for the coarse fraction, since partial dissolution of the crystals destroyed completely the ESR signal of the hexagonal phase. The thickness of the removed layer was only 4.5 μ, and, therefore, we concluded that the coarse fraction had a thinner hexagonal layer. The corrections to Δ for the inhomogeneity of the manganese distribution were included in Table 4 (they are given in parentheses).

In all the experiments described above, we used ZnS samples in which the average manganese concentration was 10^{-4} g-atom/mole. The role of the manganese diffusion in the for-

TABLE 4

Fraction	d_{av}, μ	First method		Second method	
		R	Δ, μ	R'	Δ, μ
Fine	25	0.16	10.5	0.03	9.6
Intermediate	100	2.75	5.9 (−0.5)	0.25	10.7 (+0.8)
Coarse	210	8.8	4	—	4.5

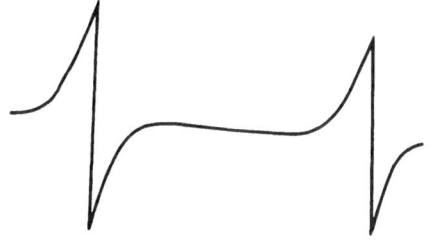

Fig. 7. ESR spectrum (low-field part) of a
ZnS:Mn powder activated during growth and
subsequently heated at T = 1170°C. The concentration of Mn in this sample was 10^{-3}
g-atom/mole.

mation of the hexagonal surface layer was determined in similar experiments using lower Mn
concentrations. Moreover, we also used melt-grown crystals [6].

The method described above was used to prepare fine and intermediate fractions of grains
with manganese concentrations amounting to 10^{-5} and 10^{-6} g-atom/mole. The ESR spectra of
these samples, recorded under identical conditions, indicated that the fraction of the Mn^{2+} centers in the hexagonal phase of ZnS in each sample decreased with decreasing manganese concentration. The Mn concentration was 10^{-5} g/g; the effective thickness of the hexagonal layer
in the intermediate and fine fractions differed considerably and was 2.5 and 4.6 μ. This indicated that the hexagonal layer had a greater fraction of the cubic phase. The values of Δ for
the samples prepared under identical activation conditions from ZnS crystals were the same
for the crystals prepared by the sublimation method and from the melt under pressure. Moreover, the effective thickness of the hexagonal layer was not affected by a change in the atmosphere (during heating) from NH_3 to H_2; moreover, the elimination of the chloride flux had no
influence. However, when manganese was introduced into the ZnS lattice at the moment of its
formation, i.e., during the growth of the original crystals, a second heating of the crushed
samples at 1170°C did not produce the hexagonal phase. This experiment was carried out on
a ZnS crystal grown from the melt and activated with manganese in a concentration of 10^{-3} g/g
during growth [6]. When the crystal was pulverized to the particle size of the fine fraction
(d_{av} = 10–15 μ), the sample obtained was heated at 1170°C for 30 min with and without flux. In
this case, the ESR spectrum (Fig. 7) indicated that the Mn^{2+} centers, typical of the hexagonal
surface layer, were no longer observed. Consequently, this layer was formed only when an
impurity was introduced into a crystal by diffusion at high temperatures.

CHAPTER II

CUBIC AND HEXAGONAL ZnS WITH Eu^{2+} IMPURITY CENTERS

Since the formation of different manganese centers in ZnS is largely governed by the lattice
structure of the host, we can expect other impurities also to form a variety of centers in ZnS.
This is manifested particularly clearly in the case of ZnS:Eu. The centers containing Eu^{2+} are
characterized by a high luminescence efficiency (when excited with near ultraviolet radiation)
so that their spectrum can be separated from the spectrum of accidental impurities and the
relationship between the various types of center and the crystal structure can be determined.
Moreover, the Eu^{2+} ions give rise to a strong ESR signal. The state of these ions is $^8S_{7/2}$. An
external magnetic field lifts the degeneracy of this state so that seven absorption transitions
between the sublevels can be observed. Each of the ESR lines has a hyperfine structure consisting of 12 components, and this structure is due to the nuclear spin I = 5/2 of the Eu^{151} and
Eu^{153} isotopes.

In contrast to the Mn^{2+} ion, no work has been done on europium impurities in zinc sulfide.
However, trigonal and cubic Eu^{2+} centers have been found in other II-VI compounds. The trig-

onal Eu^{2+} centers occur in CdS [17, 18] and CdSe [19]. The g factor is isotropic but the initial splitting of the spin level is considerable. The cubic centers occur in CdTe [19]. Hence, we may expect the ESR spectra of ZnS:Eu to indicate the presence of the cubic and trigonal Eu^{2+} centers.

§ 1.   Elementary Luminescence Bands of ZnS:Eu

The room-temperature luminescence spectrum of powdered ZnS:Eu excited with λ = 365-nm radiation has two clear maxima at hν = 2.3 and 1.8-1.9 eV. However, our preliminary experiments indicated that samples with different amounts of europium had somewhat different spectra with different positions of these maxima. Moreover, we found that the spectra were strongly affected by the temperature and wavelength of the exciting light. This demonstrated that the luminescence bands were not elementary, and the spectrum consisted of a superposition of bands representing centers of different kinds.

The relationship between the optical properties of centers of each type and the structure of the immediate environment of these centers can be found by identifying the elementary bands in the luminescence spectrum of ZnS:Eu samples with different proportions of the cubic and hexagonal phases.

We investigated a batch of phosphors prepared by heating deoxidized zinc sulfide powder containing an admixture of europium nitrate. This heating took place at 1200°C and lasted 30 min in an H_2S atmosphere. The concentration of europium in the charge was 10^{-4}, $5 \cdot 10^{-4}$, $2.5 \cdot 10^{-3}$, $5 \cdot 10^{-3}$, and 10^{-2} g/g. The phosphors prepared in this way had a predominant hexagonal structure except for the samples with the highest concentration of Eu because heating took place at a temperature exceeding the sphalerite—wurtzite phase-transition point. It was possible to prepare directly samples with the same activator concentrations but with a predominantly cubic structure because the diffusion of europium in zinc sulfide was extremely slow below the phase-transition point due to the large ionic radius of the activator ($R_{iEu^{2+}}$ = 1.20 Å). Therefore, samples with the cubic structure were prepared by applying a hydrostatic pressure [20] of about $4 \cdot 10^3$ kgf/cm^2 to a europium-activated hexagonal sample of ZnS. This treatment shifted the atomic layers in the microcrystals and produced a stable cubic close packing. A Debye diffraction pattern obtained for control samples indicated an almost complete conversion to the cubic modification. Contamination of the phosphor with metal particles of the press, which could occur at high pressures, was avoided by wrapping the sample in polyethylene. This method was used in the preparation of a batch of cubic phosphors with the same europium concentration as that in the hexagonal samples.

The luminescence spectra of ZnS:Eu samples with a predominantly hexagonal structure were determined at 77°K by excitation with ultraviolet radiation of two (mercury emission) wavelengths of λ = 313 and 365 nm. A comparison of the spectra of this batch indicated that when the activator concentration was increased the hν = 1.85 eV maximum shifted somewhat toward the long-wavelength edge. The difference between the spectra made it possible to analyze them into elementary bands by the generalized Alentsev method [21], in which no assumptions were made about the shape of the individual bands. As a result of this analysis, we established that the spectra of all the hexagonal samples consisted of the same three elementary bands with maxima at $h\nu_{max}$ = 1.75, 1.90, and 2.26 eV [20] and with widths of 0.194, 0.240, and 0.140 eV, respectively; the spectra differed only in respect of the relative intensities of these bands (Fig. 8).

A similar analysis of the spectra was also carried out for ZnS:Eu samples with a predominantly cubic structure (these samples were obtained by compressing a hexagonal phosphor). The spectra of these samples differed considerably from those of the hexagonal samples with the same europium concentrations. An analysis which yielded the elementary bands established

I_1, rel. units

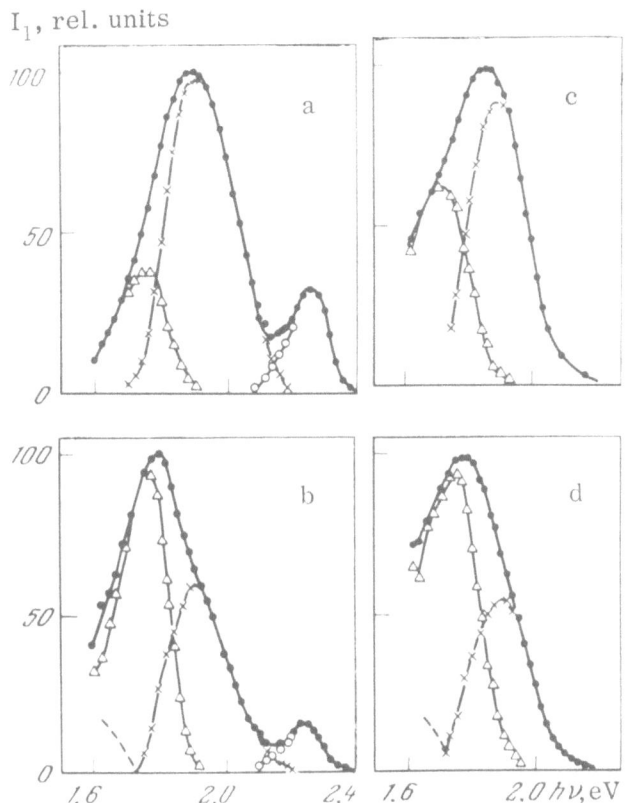

Fig. 8. Luminescence spectra and profiles of elementary bands: a, c) ZnS:Eu (5×10^{-4} g/g); b, d) ZnS:Eu (1×10^{-2} g/g); a, b) "hexagonal" sample, c, d) cubic sample.

that the characteristic features of the spectra of the cubic samples reduced to an enhancement of the $h\nu_{max}$ = 1.75 eV band compared with the $h\nu_{max}$ = 1.9 eV band and an almost complete disappearance of the $h\nu_{max}$ = 2.26 eV band.

The positions of the bands were the same in the hexagonal and cubic spectra but the half-width was somewhat greater in the cubic spectra. The bands of the hexagonal phosphors remained narrow even when they contained a considerable proportion of europium. Hence, we concluded that the band broadening was due to lattice imperfections generated by compression. Distortions in this lattice were indicated also by the broadening of the large-angle lines in the Debye diffraction pattern.

In addition to these bands, the spectra of the samples with the cubic lattice had a weak red band with a maximum at $h\nu_{max}$ < 1.58 eV. The long-wavelength edge of this band (Fig. 8) was not investigated in detail because of the low sensitivity of our spectrometer. The luminescence spectrum of a cubic phosphor prepared by hydrostatic compression was fully analogous to the spectrum of a predominantly cubic single crystal grown from the vapor phase and then diffusion-activated with europium. Consequently, the differences between the spectra of the hexagonal and cubic phosphors were due to the different lattice structures and not to the defects generated by compression.

The results obtained indicated that the europium atoms in zinc sulfide formed four types of luminescence centers, two of which ($h\nu_{max}$ = 1.75 and 1.90 eV) were present in the cubic and hexagonal samples.

§2. ESR Spectra of Europium Ions in Cubic

and Hexagonal ZnS

All the investigated cubic and hexagonal ZnS:Eu samples exhibited intense ESR spectra at low and room temperatures. The intensity distributions in these spectra depended unambiguously on the amount of europium present in the ZnS lattice. The ESR spectra of the cubic and hexagonal phosphor powders differed quite considerably. The spectrum of the hexagonal powder consisted of seven wide overlapping lines with a poorly resolved structure in the range of fields from 2500 to 6500 Oe. In contrast, the spectrum of the cubic phosphor consisted of one line with g = 1.997 ± 0.002, which also had pronounced structure (Fig. 9). The very appearance of a strong ESR spectrum demonstrated that europium was present in ZnS in the form of Eu^{2+} ions, characterized by a hyperfine structure. Estimates of the amount of europium obtained by comparing the ESR spectra of ZnS:Eu and other compounds containing paramagnetic centers ($CuSO_4 \cdot 5H_2O$, DPPH) demonstrated that practically all the europium in ZnS was present in the form of Eu^{2+}. The poor resolution of the ESR spectrum, associated with the line broadening, did not improve when the samples were cooled to 77°K and was practically independent of the europium concentration between $5 \cdot 10^{-5}$ and $5 \cdot 10^{-3}$ g/g. Consequently, the widths of the ESR lines were governed mainly by the anisotropy of the spectrum and not by the spin—lattice or spin—spin interactions.

The principal parameters of the spin Hamiltonian of the europium ions were determined from the ESR spectrum of a ZnS:Eu single crystal. As pointed out earlier, a zinc sulfide crystal with a large number of defect layers and isomorphous substituent impurities should contain both cubic and trigonal centers. The axes of the cubic and trigonal fields of the ions surrounding these centers are related to one another and have a definite orientation relative to the c crystal axis and to the crystal faces. The distribution of the axes in a hexahedral prism, typical of the grown ZnS crystals, is shown in Fig. 10. The centers in the cubic environment have threefold and fourfold symmetry axes, which are — respectively — the body diagonals and the edges of the cubic unit cell. The centers in the trigonal environment field have a symmetry axis

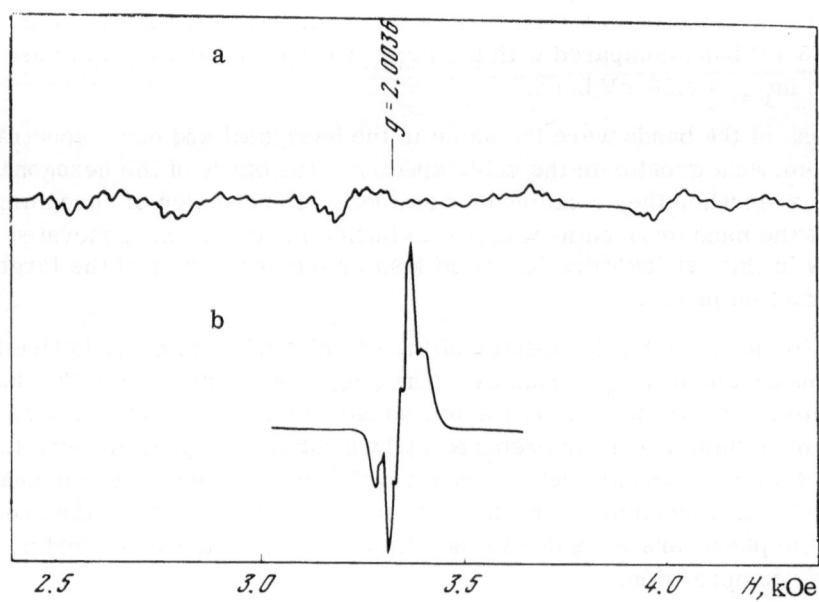

Fig. 9. ESR spectra of hexagonal (a) and cubic (b) ZnS:Eu powders.

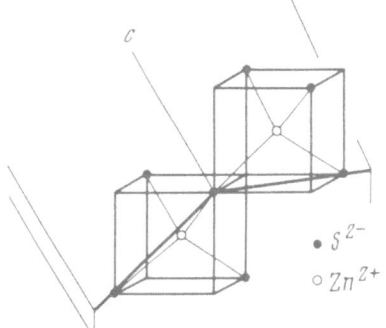

Fig. 10. Elementary tetrahedra in a hexagonal ZnS crystal.

parallel to the c crystal axis. The number of impurity ions and their distribution between the cubic and trigonal centers can be found for a specific crystal by determining the symmetry elements of the properties of the centers reflecting the structure of the environment. In the case of the paramagnetic Eu^{2+} ion in the ZnS lattice, the characteristic property of the centers is their anisotropic ESR spectrum.

Single crystals of ZnS activated with Eu during growth exhibited the ESR spectrum under the same conditions as the powder samples. This spectrum could be regarded as a superposition of a strongly anisotropic spectrum on one which was almost isotropic (Fig. 11). The anisotropic spectrum consisted of seven lines of which the strongest was located at the center and overlapped the isotropic spectrum for any orientation of the crystal. The positions of the other six lines showed an extremum when a crystal was rotated in a magnetic field. All the lines were split into 12 hyperfine components. This pattern was in agreement with the assumption that the Eu^{2+} ions were in the $^8S_{7/2}$ state and located at sites of cubic or lower symmetry.

The symmetry of the crystal field produced by the environment of the paramagnetic ions was determined by recording the dependences of the ESR spectra on the orientation of the principal crystal axes relative to the external magnetic field H. These dependences were obtained by rotating a crystal (see Fig. 10) about an axis perpendicular to the field H and were recorded for the following orientations of the crystal axes relative to the rotation axis O perpendicular to the magnetic field:

1) the body diagonal [111] of an elementary cube coinciding with the c crystal axis perpendicular to the rotation axis;

2) the body diagonal of an elementary cube coinciding with the c crystal axis and parallel to the rotation axis.

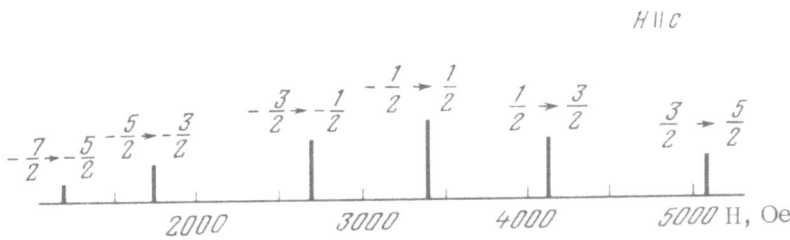

Fig. 11. Schematic representation of the ESR spectrum of a ZnS:Eu single crystal. The positions of the centers are given for six groups of hyperfine-structure lines.

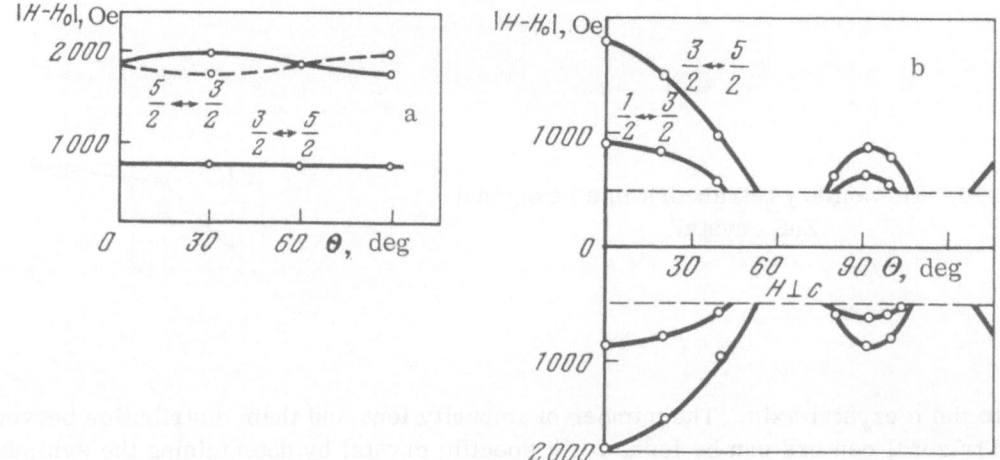

Fig. 12. Angular dependence of the ESR spectrum of ZnS:Eu due to rotation of a crystal about the O axis: a) O ∥ c; b) O ⊥ c; the measurements were not carried out within the region bounded by the dashed lines in Fig. 12b.

The dependences of the positions of the lines in the anisotropic spectrum on the angle of rotation of a crystal are plotted in Fig. 12 for these two orientations. In the H ∥ c case, the extremal positions of the lines with the largest relative displacement is obtained. In the H ⊥ c case, once again the line positions are extremal but with the smallest relative displacement. A comparison of these dependences shows that the anisotropic spectrum is due to centers with axial symmetry. The $M = {}^5/_2 \leftrightarrow {}^3/_2$ transition line is a doublet (Fig. 12a).

The ESR spectrum of an Eu^{2+} ion in a crystal field of axial symmetry can be described by the Hamiltonian [17]

$$\mathcal{H} = g\beta \left[\mathbf{H}.\mathbf{S} + \frac{1}{3} b_2^0 O_2^0 + \frac{1}{60} b_4^0 O_4^0 + \frac{1}{1260} b_6^0 O_6^0 + A\mathbf{S}.\mathbf{I}\right], \qquad (2.1)$$

where g, b_2^0, b_4^0, b_6^0, and A are the experimentally determined parameters; S = 7/2, I = 5/2; the functions O_m^0 are given in [2]. In the H ∥ c case, the eigenvalues of the Hamiltonian describe the following levels $E_{\pm M}$, deduced ignoring the hyperfine interaction

$$\left.\begin{aligned}
E_{\pm 7/2} &= \pm \frac{7}{2} g\beta H + g\beta (7b_2^0 + 7b_4^0 + 6b_6^0), \\
E_{\pm 5/2} &= \pm \frac{5}{2} g\beta H + g\beta (b_2^0 - 13b_4^0 - 5b_6^0), \\
E_{\pm 3/2} &= \pm \frac{3}{2} g\beta H + g\beta (-3b_2^0 - 3b_4^0 + 9b_6^0), \\
E_{\pm 1/2} &= \pm \frac{1}{2} g\beta H + g\beta (-5b_2^0 + 9b_4^0 - 5b_6^0).
\end{aligned}\right\} \qquad (2.2)$$

The hyperfine interaction described by the term $A\mathbf{S}.\mathbf{I}$ in Eq. (2.1) gives rise to an additional displacement of these levels by

$$E'_{M, m} = g\beta A M m + \frac{(g\beta A)^2}{2h\nu} \{M [I (I + 1) - m^2] - m [S (S + 1) - M^2]\}. \qquad (2.3)$$

When this displacement is allowed for, the resonance fields H_M for the $M - 1 \rightarrow M$ transitions are given by

$$H_{7/2} = H_0 - 6b_2^0 - 20b_4^0 - 6b_6^0 - Am - \frac{A^2}{2H_0}\left(\frac{35}{4} - m^2 + 6m\right),$$

$$H_{5/2} = H_0 - 4b_2^0 + 10b_4^0 + 14b_6^0 - Am - \frac{A^2}{2H_0}\left(\frac{35}{4} - m^2 + 4m\right),$$

$$H_{3/2} = H_0 - 2b_2^0 + 12b_4^0 - 14b_6^0 - Am - \frac{A^2}{2H_0}\left(\frac{35}{4} - m^2 + 2m\right),$$

$$H_{1/2} = H_0 - Am - \frac{A^2}{2H_0}\left(\frac{35}{4} - m^2\right), \qquad\qquad\qquad\qquad\qquad (2.4)$$

$$H_{-1/2} = H_0 + 2b_2^0 - 12b_4^0 + 14b_6^0 - Am - \frac{A^2}{2H_0}\left(\frac{35}{4} - m^2 - 2m\right),$$

$$H_{-3/2} = H_0 + 4b_2^0 - 10b_4^0 - 14b_6^0 - Am - \frac{A^2}{2H_0}\left(\frac{35}{4} - m^2 - 4m\right),$$

$$H_{-5/2} = H_0 + 6b_2^0 + 20b_4^0 + 6b_6^0 - Am - \frac{A^2}{2H_0}\left(\frac{35}{4} - m^2 - 6m\right)$$

If the resonance field of each fine-structure line is assumed to be the average of the fields corresponding to the hyperfine components, the separations δH_M between the lines with different values of M, on the one hand, and the line with M = 1/2, on the other, are:

$$|\delta H_{7/2}| = 6b_2^0 + 20b_4^0 + 6b_6^0 = P,$$

$$|\delta H_{5/2}| = 4b_2^0 - 10b_4^0 - 14b_6^0 = Q, \qquad\qquad\qquad (2.5)$$

$$|\delta H_{3/2}| = 2b_2^0 - 12b_4^0 + 14b_6^0 = R.$$

Hence, it follows that

$$b_2^0 = \frac{1}{84}(7P + 8Q + 5R), \qquad b_4^0 = \frac{1}{308}(7P - 6Q - 9R),$$

$$b_6^0 = \frac{1}{132}(P - 4Q + 5R). \qquad\qquad\qquad (2.6)$$

The parameters b_n^0, deduced from the experimental values of the field intervals (separations) between the fine-structure line in accordance with Eq. (2.6), are listed in Table 5 (the sign of b_2^0 was assumed to be positive) together with the values of the g factor and the constant A for the Eu^{151} and Eu^{153} isotopes deduced from Eq. (1.23).

The relatively large value of b_2^0 shows that the axial crystal field exerted on the investigated Eu^{2+} centers is high, and these centers are located in the hexagonal part of a crystal.

We were unable to determine the parameters of the centers responsible for the weakly anisotropic ESR spectrum in the g = 2.0 region because of the superposition of several structure lines.

§ 3. Eu^{2+} Centers in ZnS Mixed-Structure Powders

A comparison of the ESR spectra of ZnS:Eu powders with the hexagonal structure and of a single crystal with a mixed structure indicated that these spectra occurred in practically

TABLE 5

Sample	b_2^0, Oe	b_4^0, Oe	b_6^0, Oe	g	A^{151}	A^{153}
ZnS-Eu						
Hexagonal	390	—11	~1	1,980	28	12
CdS-Eu [17]	349.9	—12.4	0.7	1,992	24.1	10.8

the same range of magnetic fields. Although the lines in the former spectrum were not satisfactorily resolved, we were still able to conclude that they were the anisotropically broadened
fine-structure lines due to the Eu^{2+} centers of the same kind as in the hexagonal phase of a
single crystal or due to centers with similar parameters. This enabled us to determine quantitatively the proportions of the crystal phases in the ZnS:Eu phosphor from the ESR spectra.
We deduced from these spectra that the proportion of the cubic phase increased with increasing
concentration of europium in "high-temperature" phosphors, i.e., when the phosphor was
heated at 1200°C. The proportion of the cubic phase reached 70% (curve 4 in Fig. 13) when the
activator concentration was 10^{-2} g/g. In this case, the ESR spectrum contained lines typical
of the hexagonal and cubic ZnS lattices. The appearance of the cubic phase at high europium
concentrations was confirmed by the x-ray structure analysis data. The Debye diffraction
pattern of the phosphor with 10^{-2} g/g europium had a spectrum in which the strongest lines
were due to the cubic lattice. This was due to the fact that the compound EuS formed in the
phosphor had the cubic structure with the lattice constant ($a = 5.7$ Å) close to that of ZnS. This
compound initiated the formation of the sphalerite phase.

The increase in the cubic phase content with rising europium concentration in the high-
temperature ("hexagonal") phosphors had the dominant effect on the concentration dependences
of the luminescence spectra of these phosphors in the $h\nu = 1.6$–2.0 eV range. This was manifested particularly by the fact that the spectrum of a "high-temperature" phosphor containing
10^{-2} g/g europium exhibited a band at $h\nu_{max} < 1.58$ eV, characteristic of the cubic phase alone
(Fig. 8). The quantitative information on the cubic-phase content deduced from the ESR spectra made it possible to eliminate the contribution of the cubic centers to the spectra of the
"high-temperature" phosphors which were generally mixtures of the cubic and hexagonal modifications. The luminescence spectra were characterized by the ratio of the intensities of the
bands with maxima at 1.75 and 1.90 eV. The concentration dependences of these ratios were
plotted in Fig. 13 for a cubic phosphor (curve 1) and a predominantly hexagonal phosphor
(curve 2). Such elimination made it possible to obtain the value of this ratio for the luminescence spectra of the purely hexagonal phase with different europium concentrations (curve 3).
It is clear from Fig. 13 that this ratio was practically independent of the europium concentration in the range $5 \cdot 10^{-4}$–$5 \cdot 10^{-3}$ g/g, corresponding to the most efficient ZnS:Eu phosphors.

The ESR spectra of the Eu^{2+} centers responsible for the bands with maxima at 1.75 and
1.90 eV were not resolved because the differences between their parameters were slight.

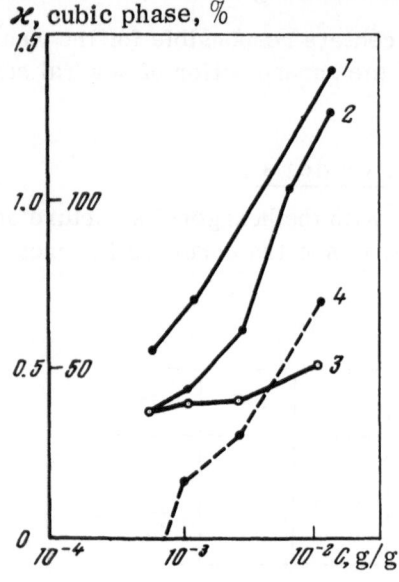

Fig. 13. Influence of the lattice structure on the
luminescence spectra: 1) ratio ($\varkappa$) of the intensities of the bands with maxima at 1.75 and
1.90 eV emitted by a cubic phosphor; 2) corresponding ratio for a "hexagonal" phosphor; 3)
pure hexagonal phase; 4) fraction of the cubic
phase in a "hexagonal" phosphor deduced by the
ESR method.

CHAPTER III

INFLUENCE OF IRRADIATION ON
Cr^{3+} CENTERS IN RUBY

Color centers appear in ruby crystals as a result of bombardment with gamma and x rays. Similar centers can be produced in ruby by high-power optical excitation (for example, using pulses produced by a xenon lamp).

Several investigators [22, 23] have attributed the formation of these centers to the transfer of the Cr^{3+} ions, replacing isomorphously the Al^{3+} ions in the corundum lattice, to the Cr^{2+} and Cr^{4+} states. This is in agreement with the observation that when the gamma-ray dose is sufficiently high the concentration of the Cr^{3+} ions, deduced from the ESR spectrum [24], decreases by more than 40%. In addition to the centers associated with the chromium ions, hard irradiation produces centers associated with intrinsic structure defects. One of these centers is an aluminum vacancy which has captured (as a result of irradiation) one hole. Such a vacancy is paramagnetic, and it has been found in fairly pure corundum crystals [25]. The new centers which appear in ruby not only alter considerably the optical absorption of the crystal as a whole, but also distort the local crystal field near some of the main Cr^{3+} centers. Such distortions may alter the luminescence and absorption associated with these centers. Since the luminescence and absorption are among the most important properties of ruby crystal phosphors, it is desirable to investigate in greater detail the processes involving charge exchange between the centers under the influence of optical and hard radiation. In particular, it is desirable to know which of the additional absorption bands are due to the Cr^{2+} and Cr^{4+} ions, and to estimate the influence of the electric fields generated by the centers which have experienced charge exchange on the properties of the main Cr^{3+} ions.

§ 1. Behavior of Cr^{3+} Ions as Electron and Hole Traps

The Cr^{3+} ions in the ruby lattice replace the Al^{3+} in such a way that they become surrounded by six O^{2-} ions forming an octahedron. This octahedral environment produces a strong crystal electric field which has a considerable influence on the energy levels of the Cr^{3+} ion. The Al^{3+} ions located further from Cr^{3+} do not affect significantly the Cr^{3+} energy levels. They simply broaden the ESR lines because of the magnetic dipole interactions of the moments of the Al^{27} nuclei with the electron magnetic moment of Cr^{3+}. There are four different positions that the Cr^{3+} ions can occupy in the ruby lattice. All these positions are energetically equivalent if a crystal is not subjected to an external electric field. The crystal field in ruby acting on the Cr^{3+} ions is entirely trigonal because these ions are not located at the centers of the octahedra formed by the O^{2-} ions.

The Cr^{3+} ion has three electrons in the partly filled 3d shell. Consequently, the ground state of the free ion is $^4F_{3/2}$, and the electron spin is S = 3/2. The axial crystal field splits the ^4F level so that the lower sublevel is the orbital singlet L = 0 which is quadruply (2S + 1) spin-degenerate. The same axial field splits additionally the quadruply degenerate sublevel into two doubly degenerate spin levels separated by an energy gap of 0.38 cm^{-1}. The splitting of the orbital level is very large compared with this gap and the transitions between the lower and upper split components lie in the optical part of the spectrum (U and Y absorption bands of ruby). The transition from metastable levels of the doublet system to the lower ^4A level corresponds to the R-line luminescence. The application of a magnetic field lifts the remaining double degeneracy of the spin levels, and the absorption transitions between the split levels are observed in ESR.

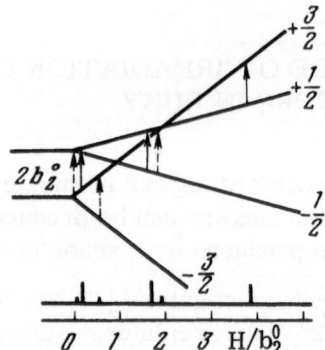

Fig. 14. Energy levels of the ground state of Cr^{3+} ions in corundum subjected to H ∥ c [9].

The ESR spectrum recorded subject to the condition $h\nu > g\beta b_2^0$ (b_2^0 is the constant representing the initial splitting of the spin levels by the axial crystal field) for c ∥ H has three lines corresponding to magnetic fields $H_1 = H_0 - |2b_2^0|$, $H_2 = H_0$, and $H_3 = H_0 + |2b_2^0|$ for the m = $^3/_2 \rightarrow ^1/_2$, $M = -^1/_2 \rightarrow ^1/_2$, and $M = ^1/_2 \rightarrow ^3/_2$ transitions, respectively (Fig. 14). (If $H < |2b_2^0|$, the M = 3/2 level is located below the M = 1/2 level.) The constants [26] are g = 1.989 and $b_2^0 = 2043$ Oe.

The Cr^{2+} and Cr^{4+} ions have an even number of electrons but are also paramagnetic. The ESR method can be used, in principle, to detect these ions in irradiated ruby. The Cr^{4+} ion has two electrons in the 3d shell, and the total spin of these electrons is S = 1. The trigonal field, due to the distortion of the octahedron of six oxygen ions containing the Cr^{4+} ion, splits — by the spin—orbit interaction — the ground state (triplet) into a singlet and doublet [27]. A small admixture of a field of lower symmetry gives rise to a further splitting of the doublet M = ±1. The ESR spectrum of Cr^{4+} in ruby, which unavoidably (because of the growth conditions) contained such ions ("orange ruby"), was investigated in the temperature range 2-77°K. When the temperature was raised from 2°K, the intensity of the only observed line increased and passed through a maximum at 4.2°K. At higher temperatures, the width of the line rose rapidly, and the ESR signal disappeared completely near 77°K. The strong temperature dependence of the line intensity in the 2-4.2°K range indicated that the upper sublevel was a split doublet, and the ESR line was due to the $\Delta M = 2$ transition between the components of the doublet, forbidden in the absence of a low-symmetry field.

The observed rapid broadening of the line with increasing temperature was due to the strong temperature dependence of the spin—lattice relaxation time, which was less than 10^{-9} sec below 77°K. Therefore, the ESR line intensity did not become saturated at 4.2°K even when the microwave power in the spectrometer resonator was increased to 0.1 W. This line was also observed in the ESR spectrum of the ordinary ("rose") ruby irradiated with x rays [28].

The Cr^{2+} ion has four electrons (state $^5D_0$) in the partly filled 3d shell. In the cubic (octahedral or tetrahedral) field, the ground state of these ions, which has quintuple (2L + 1) orbital degeneracy, is split into a triplet and doublet, the latter becoming the lower orbital level [29]. The trigonal field splits the doublet so that the ground state has no orbital degeneracy. All the states, including the ground, have quintuple (2S + 1) spin degeneracy. Since the Kramers theorem is inapplicable to the even number of electrons, the spin—orbit coupling probably also lifts the spin degeneracy, and five singlet levels are formed even in the absence of an external magnetic field. In general, the separation between these levels can be very large, and the ESR may not be observed in the centimeter wavelength range. Experimental investigations of the ESR of Cr^{2+} ions established its presence in $CrSO_4 \cdot 5H_2O$ in the millimeter wavelength range and in chromium-doped CdS (hexagonal lattice) in the centimeter range. In both cases, it was necessary to cool the samples to liquid helium temperature.

The ESR spectrum of the Cr^{2+} ion in the Al_2O_3 lattice was not observed, and, therefore, the presence of this ion in ruby had to be deduced by other methods. We used the visible absorption bands which appeared as a result of bombardment of ruby with gamma rays. The relationship between the chromium ions and the color centers was investigated by determining the spectral dependence of the influence of optical radiation on the ESR spectrum of Cr^{3+} in irradiated ruby samples.

A batch of samples with 0.01, 0.03, and 0.1% chromium concentrations was bombarded with gamma rays in a dose of 10^6 rad using a Co^{60} source. A comparison of the amplitudes of the ESR lines of the Cr^{3+} ions before and after irradiation indicated that the intensity of the ESR spectrum was reduced considerably by the irradiation.

This reduction in the paramagnetic absorption at the maximum of the ESR line as a result of irradiation with gamma rays was not accompanied by changes in the line profile and half-width, so that we concluded that the weakening of the ESR spectrum was due to a reduction in the area under the line and was caused by the transfer of some of the Cr^{3+} ions to a different valence state, i.e., it was due to a reduction in the concentration of these ions in the sample.

We assumed that the intensity of the $M = -\frac{1}{2} \to \frac{1}{2}$ line of our samples was unity, and we then found that the intensities of the lines emitted by the irradiated samples with 0.01, 0.05, and 0.1% chromium concentrations were, respectively, 0.82 ± 0.03, 0.84 ± 0.02, and 0.80 ± 0.04.

In addition to this effect in the gamma-irradiated sample (0.05% Cr), we observed an ESR spectrum with one anisotropic line of somewhat asymmetric profile only near 4.2°K (Fig. 15). The resonance field was $H_{res} \approx 1900$ Oe for $H \parallel c$ when the spectrometer frequency was $\nu = 9480$ MHz. Cooling to 77°K gradually weakened this line which eventually disappeared. The maximum intensity (amplitude) of this line was approximately an order of magnitude lower than the intensity of the $M = -\frac{1}{2} \to \frac{1}{2}$ line of the Cr^{3+} ion. The absence of this line before irradiation, the asymmetric profile, and the characteristic temperature dependence suggested that it should be attributed to the Cr^{4+} ions observed earlier in orange ruby [27].

The reduction in the intensity of the ESR lines of Cr^{3+} after bombardment with gamma rays and the appearance of the new line were accompanied by changes in the optical absorption spectrum. This spectrum acquired a series of overlapping bands. In the 200-600 nm range, we found bands with maxima at $\lambda = 540$, 470, and 260-330 nm, as well as at $\lambda < 260$ nm [30, 31]. It was assumed that these maxima coincided approximately with the maxima of the individual additional absorption bands. Illumination of a gamma-irradiated sample with monochromatic light of sufficient intensity and of wavelengths coinciding with the maxima of some of these bands enhanced the intensity of the ESR spectrum of the Cr^{3+} ions until it recovered its initial value before the irradiation with gamma rays. This was observed when the gamma-irradiated ruby samples were illuminated with monochromatic light of $\lambda = 550$ and 400 nm wavelength whereas light of $\lambda = 470$ nm wavelength had no effect on the ESR spectrum.

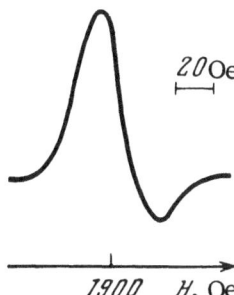

Fig. 15. ESR line of Cr^{4+} in gamma-irradiated ruby.

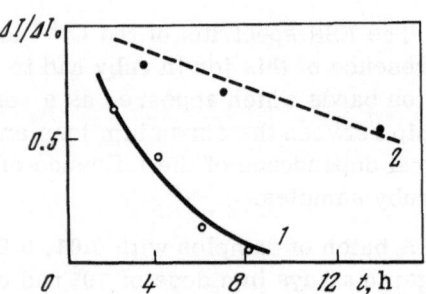

Fig. 16. Recovery of the Cr^{3+} ion concentration in gamma-irradiated ruby as a result of illumination with light of different wavelengths: 1) $\lambda = 365$ nm; 2) $\lambda = 550$ nm.

The efficiency of the optical effect in these "active" bands was quite different. The strongest effect on the ESR intensity was observed on illumination with light whose maximum was located in the vicinity of $\lambda = 400$ nm. For example, illumination with $\lambda = 365$ nm light for 10-12 h practically restored the initial intensity of the ESR spectrum of the Cr^{3+} ions in gamma-irradiated ruby. The action of light in a band with a maximum at $\lambda = 550$ nm was 6-8 times weaker, and, therefore, complete recovery of the ESR spectrum was not achieved.

The results of optical illumination with light of these wavelengths on the gamma-irradiated ruby with 0.05% chromium are plotted in Fig. 16. Here, the abscissa shows the duration of illumination and the ordinate, the ratio $\Delta I / \Delta I_0$, where ΔI is the change in the ESR intensity caused by the absorbed light quanta, and ΔI_0 is the change in intensity caused by gamma irradiation. Similar results were also obtained for a gamma-irradiated ruby sample with 0.1% Cr. The only difference was a somewhat longer (by a factor of ~1.5) exposure needed to restore the ESR spectrum.

Heating of the gamma-irradiated samples to 350°C also restored the number of Cr^{3+} ions present before irradiation. Figure 17 shows the dependence of the change in the density of the ESR $M = -1/2 \rightarrow 1/2$ line emitted by a gamma-irradiated sample (0.05% Cr) on the annealing temperature. The experimental points were obtained by heating at increasing temperatures (5 min at each temperature) and rapid cooling to room temperature at which the ESR intensity was measured. It is clear from this dependence that a significant annealing of the effects of gamma irradiation was completed basically at about 300°C. The line intensity was totally restored by heating to at least 500°C, when the concentration of the Cr^{3+} ions recovered the value before the irradiation. This treatment also destroyed completely the ESR line attributed to the Cr^{4+} ions ($H_{res} \approx 1900$ Oe).

All these properties of the gamma-irradiated ruby indicated that the interaction of relatively energetic quanta with ruby produced at least two types of center associated with chromium, but differing in respect of their paramagnetic and optical characteristics. If the centers absorbing light in the 200-600-nm range were the Cr^{4+} and Cr^{2+} ions, the "active" bands, i.e., those additional absorption bands which corresponded to the illumination that altered the

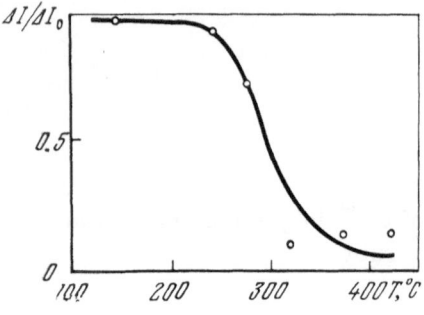

Fig. 17. Recovery of the Cr^{3+} ion concentration in gamma-irradiated ruby as a result of annealing.

number of Cr^{3+} ions in the gamma-irradiated crystals, should be attributed to the Cr^{2+} ions which acted as filled electron traps. The other additional absorption bands at which there were no changes in the ESR spectra as a result of illumination with light corresponding to these bands should be attributed to the intracenter $a^3T_1 \rightarrow {}^3T_2(t_2e)$ transitions in the Cr^{4+} ions [25, 31] which did not alter the charge state of these ions.

It follows from the energy-level scheme of chromium ions in the forbidden band of corundum [31] that the illumination of gamma-irradiated ruby, in which the Cr^{2+} ionic charge is compensated mainly by the Cr^{4+} ions, with monochromatic light of wavelengths corresponding to the maximum of any one of the active bands transfers electrons from the Cr^{2+} ions to the conduction band, and these electrons then recombine with holes at the Cr^{4+} ions. Thus, one absorption of a "deexciting" photon restores simultaneously two Cr^{3+} ions.

According to the published information, the Cr^{2+} ion has two excited levels in the forbidden band of corundum. The higher level is located near the bottom of the conduction band and corresponds to the optical absorption at $\lambda_{max} = 400$ nm. The transfer of an electron from this level to the conduction bands is more likely, which is indicated by the relatively high efficiency of the illumination with light of wavelength $\lambda = 365$ nm in the process of conversion of the Cr^{2+} ions to the charge state Cr^{3+}. The lower excited level of Cr^{2+}, which corresponds to the absorption band $\lambda_{max} = 550$ nm, lies deeper in the forbidden band, and, therefore, a direct liberation of an electron from this level to the conduction band is unlikely. However, our experimental results show that such transitions do occur, since illumination with light in the $\lambda_{max} = 550$ nm band results in a considerable reduction in the number of Cr^{2+} centers in gamma-irradiated ruby. It is shown in [32] that the conduction band of corundum (ruby) is bent considerably. The bottom of this band at k > 0 drops approximately by 2.1 eV relative to the position at k = 0. Bearing this in mind, we can attribute the "activity" of the $\lambda_{max} = 550$ nm band to the indirect transitions of electrons from the lower excited level of the Cr^{2+} ion to the conduction band. This process explains the low efficiency of the illumination with the $\lambda_{max} = 550$ nm light, which is much less than the efficiency of the illumination with $\lambda_{max} = 400$ nm.

It should be pointed out that illumination of colored crystals with light of wavelengths 330-430 and 550 nm, coinciding with the active bands, results in bleaching [31]. The heating of a gamma-colored sample to 330°C also causes bleaching in the spectral region $\lambda = 200-600$ nm. Such heating also produces a strong thermoluminescence peak with the R-line spectrum [33]. The position of this peak on the temperature scale corresponds to a depth $\varepsilon = 1.5$ eV. These results are in satisfactory agreement with our study of the influence of illumination and heating on the ESR spectrum of gamma-radiated ruby and confirm our hypothesis that charge exchange between the Cr^{3+} ions as a result of hard and high-power optical irradiation of ruby produces mainly the Cr^{2+} and Cr^{4+} ions responsible for the color centers.

§2. Distortion of Crystal Field in

Gamma-Irradiated Ruby

The ESR line width in the spectra emitted from crystals with few defects is governed mainly by the electron spin—spin interactions between the Cr^{3+} ions and electron—nuclear spin—spin interactions between the Cr^{3+} ions and Al^{27} nuclei. The experimental results show that the relative contribution of the spin—lattice interactions to the width of the lines emitted by ruby is small. The spin—lattice relaxation time, which is $0.09 \cdot 10^{-6}$ sec for ruby, gives rise to a contribution of 0.15 Oe to the line width for the samples with the lowest chromium concentration even at room temperature [26, 34].

In the case of real ruby single crystals, the line width also depends strongly on the inhomogeneity of the crystal field distribution. This inhomogeneity gives rise to a scatter of the initial splitting of the levels of the Cr^{3+} ions located at different points in a crystal. Since the positions of the $M = {}^3/_2 \to {}^1/_2$ and $M = {}^1/_2 \to {}^3/_2$ lines in the spectrum depend on the Hamiltonian constant b_2^0, the presence of regions with different values of this constant gives rise to some broadening of the lines. According to [34], the average scatter of b_2^0 in crystals not subjected to special heat treatment is about 0.1%, but annealing can reduce it to 0.001%.

A definite contribution to the ESR line broadening can also be made by a different crystal field inhomogeneity. Crystals grown by the Verneuil furnace may consist of regions with slightly different orientations of the crystal axis. The result is an anisotropic line broadening which is different for the $M = {}^3/_2 \to {}^1/_2$ and $M = {}^1/_2 \to {}^3/_2$ lines because the angular dependences of the positions of these lines on the magnetic field scale are quite different [26].

The simultaneous influence of these two sources of line broadening is responsible for the observed ESR line width $\Delta H = 14-17$ Oe of our ruby crystals.

It is reported in [28, 35] that gamma irradiation reduces the total intensity of the ESR lines and alters the widths of some of the lines. The lines associated with the transitions between the $M = 3/2$ and $M = 1/2$ levels, belonging to two different spin doublets, broaden considerably (see Fig. 14). We found that the line due to the $M = -{}^1/_2 \to {}^1/_2$ transition line was due to a change in the initial splitting of the ground level of Cr^{3+} by the crystal field, i.e., a change in the constant b_2^0. Consequently, this change should be of electrical origin. The role of the magnetic interaction of the chromium ion with the surrounding lattice ions did not cause additional broadening because this interaction would have altered the widths of all the ESR lines.

Local crystal fields could be distorted by, inter alia, the appearance of lattice defects such as interstitial ions or vacancies. These defects were indeed produced by irradiation of corundum and ruby with large reactor doses. When the dose of 1.6-MeV gamma rays was relatively small (10^4-10^6 rad), we could not expect the accumulation of large numbers of such structure defects. Therefore, the most probable cause of the line broadening was the ionization which altered the charge distributions. This would give rise to local electric fields which could change the initial splitting of the Cr^{3+} levels. Such an inhomogeneity of the crystal field in ruby also altered the profiles of the R lines in the visible range after high-power optical or gamma irradiation [36].

Quantitative estimates of the crystal field distortion due to localized charges can be obtained by considering an effective additional electric field of axial symmetry but with a randomly oriented axis. The action of an external axial electric field alters the structure of the ESR spectral lines [37]. This is due to the absence of an inversion center of the crystal field relative to the Cr^{3+} ions. Therefore, in an electric field, the energy levels of these ions are displaced, and this causes a corresponding shift of the resonance field. Since there are four energetically equivalent types of Cr^{3+} ion in the corundum lattice and these are displaced in pairs relative to the centers of the octahedra in opposite directions along the c axis, the shifts of the levels should have opposite signs, and if the applied field is sufficiently strong, the ESR lines should be split. In relatively weak fields, the profiles of these lines should change without a significant change of the area under the lines. The splitting is different for lines belonging to different transitions. In the $c \parallel H$ orientation, the application of an external electric field causes considerable changes in the $M = {}^3/_2 \to {}^1/_2$ and $M = {}^1/_2 \to {}^3/_2$ lines but no changes in the $M = -{}^1/_2 \to {}^1/_2$ line. The splitting of the first two lines can be represented quantitatively by

$$\delta H_E = 4kE_0, \tag{3.1}$$

where E_0 is the applied electric field. The constant in the above equation is $k = \partial b_2^0/\partial E = (0.93 \pm 0.01) \times 10^{-4}$ Oe·cm·V^{-1} [37].

The experimentally observed broadening of the $M = 3/2 \to 1/2$ transition lines is 12-13% for a sample of ruby with 0.01% chromium irradiated with a dose of 10^6 rad. Irradiation alters somewhat the line profiles which remain intermediate between the Gaussian and Lorentzian forms. Changes in the line parameters (i.e., the half-width ΔH and the normalized ordinate corresponding to a field $H = H_{res} - \Delta H/2$) make it possible to estimate the effective additional axial field with a random distribution of its axis. The influence of the field of an elementary charge on a Cr^{3+} ion at right angles to the crystal axis causes no line splitting because the Cr^{3+} ions of all types are displaced identically relative to the center of the unit cell. However, the action of the field is strongest along the c axis so that, in the first approximation, we may assume that the effect is entirely due to the component directed along the c axis and, consequently, $E = E_0 \cos \theta$, where θ is the angle between the axis of the additional field and the crystal axis.

The number of the Cr^{3+} centers for which the directions of the additional field axis are enclosed within an angle θ is

$$dn = n_0 \sin \theta \, d\theta. \tag{3.2}$$

If we assume a Lorentzian ESR line profile for an unirradiated sample, the net influence of charges on a Cr^{3+} ion in an irradiated sample produces a line whose profile can be described by

$$Y(H) = Y_0 \int_0^{\pi/2} \left\{ \left[1 + \left(\frac{H - H_{res} - H_{max} \cos \theta}{{}^1\!/_2 \, \Delta H} \right)^2 \right]^{-1} + \right.$$
$$\left. + \left[1 + \left(\frac{H - H_{res} + H_{max} \cos \theta}{{}^1\!/_2 \, \Delta H} \right)^2 \right]^{-1} \right\} \sin \theta \, d\theta, \tag{3.3}$$

where H_{max} is the maximum displacement of the resonance line under the action of the charge field; ΔH is the initial line half-width; H_{res} is the resonance field in the absence of charges. Calculations based on this expression yield the values of the ordinates for each value of the field H relative to the absorption maximum:

$$\frac{Y}{Y_0} = \frac{\arctan(b + a) - \arctan(b - a)}{2 \arctan a}, \tag{3.4}$$

where $b = (H - H_{res})/(\Delta H/2)$ and $a = H_{max}/(\Delta H/2)$. In this case, the half-width of the new line $\Delta H'$ is

$$\Delta H' = \Delta H \sqrt{1 + \left(\frac{2 H_{max}}{\Delta H} \right)^2}. \tag{3.5}$$

Since the displacement of the resonance line of Cr^{3+} under the influence of an external electric field E_0, oriented so that $E_0 \parallel c$, is equal to half the splitting δH_E, it follows that $H_{max} = \delta H_E/2$, and in the case of the $M = {}^3\!/_2 \to {}^1\!/_2$ transition we find from Eq. (3.1) that

$$E_0 = H/2k. \tag{3.6}$$

Thus, knowing the experimental value of the line broadening of this transition after irradiation, we can utilize Eqs. (3.5) and (3.6) to find the effective additional crystal field E':

$$E' = \left[\Delta H \sqrt{\left(\frac{\Delta H'}{\Delta H} \right)^2 - 1} \right] \Big/ 4k. \tag{3.7}$$

If the line profile for an unirradiated sample is Gaussian, the net effect of charged centers formed as a result of gamma-ray irradiation is to produce an ESR line of the Cr^{3+} ions whose profile is given by the function

$$Y(H) = Y_0 \int_0^{\pi/2} \left\{ \exp\left[-0.693 \left(\frac{H - H_{max} \cos\theta}{1/2\,\Delta H} \right)^2 \right] + \right.$$

$$\left. + \exp\left[-0.693 \left(\frac{H + H_{max} \cos\theta}{1/2\,\Delta H} \right)^2 \right] \right\} \sin\theta \, d\theta. \tag{3.8}$$

The values of this function corresponding to $H = \Delta H/2$ (Y_1) and $H = 0$ (Y_2) were obtained for different values of H_{max} by summing the integrand corresponding to 10 values of the angle θ. Thus, when Eq. (3.8) was used to find the field of the irradiation-generated centers, we had to measure not the line broadening, but the change in the ordinate at a fixed value of the magnetic field.

We investigated samples of ruby with chromium concentrations 0.01-0.1%. The samples with the lowest chromium concentration had an ESR spectrum with a nearly Lorentzian line profile. The lines of a sample with 0.1% chromium had a profile intermediate between Lorentzian and Gaussian. The width of the lines due to the $M = {}^3/_2 \rightarrow {}^1/_2$ and $M = {}^1/_2 \rightarrow {}^3/_2$ transitions, found by averaging the results of six measurements, was approximately the same for each sample. Hence, we concluded that the "mosaic effect" resulting in an anisotropic line broadening was of little significance. Irradiation with Co^{60} gamma rays (10^6 rad) increased considerably the widths of the lines due to the two transitions just mentioned (Table 6), but had practically no effect on the line due to the $M = {}^1/_2 \rightarrow -{}^1/_2$ transition. Figure 18 shows the $M = {}^3/_2 \rightarrow {}^1/_2$ line before the irradiation and the corresponding Lorentzian curve and the line after irradiation of a sample with 0.01% chromium. The same figure includes the theoretical line profile of half-width equal to that of the line exhibited by the irradiated sample but calculated in accordance with Eq. (3.8).

A similar theoretical shape of the broadened line was determined also for a sample with 0.03% chromium which had a line of half-width somewhat greater than that observed for the sample with 0.01% chromium. A satisfactory agreement between the experimental points and the calculated curves (Fig. 18) confirmed the validity of the adopted explanation of the additional broadening. The experimentally determined half-width of the lines emitted by the irradiated samples was used in estimating the effective additional fields E' acting on the Cr^{3+} ions in ruby irradiated with gamma rays. The results of estimates of this field obtained for our samples in accordance with Eq. (3.7) are listed in Table 6.

The profile of the ESR line emitted by a ruby sample with 0.1% chromium was described less satisfactorily by a Lorentzian curve. Therefore, in estimating the field E' this line was

Fig. 18. Broadening of the $M = {}^3/_2 \rightarrow {}^1/_2$ line emitted by gamma-irradiated ruby: 1) before irradiation; 2) after irradiation; 3) theoretical line profile with half-width of the irradiated sample; the dashed curve is a Lorentzian profile.

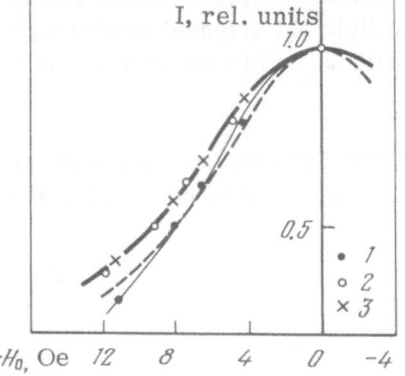

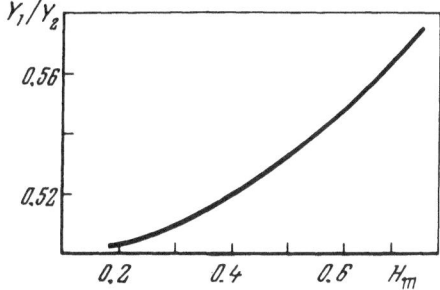

Fig 19. Dependence of Y_1/Y_2 on H_{max}.

TABLE 6

Concentration of Cr^{3+}, %	Half-width of M = $^3/_2 \rightarrow ^1/_2$ line, Oe		Broadening, Oe	Effective additional field $\times 10^{-4}$, V/cm	
	before γ irradiation	after γ irradiation		based on Lorentzian curve	based on Gaussian curve
0,01	16.4	18.3	1.9±0.2	2.2	—
0,03	18.3	20.9	2.6±0.3	2.7	—
0,1	19.2	22.3	3.1±0.3	3.0	3.4

approximated by Lorentzian and Gaussian curves. In the latter case, we started not with the half-width for an irradiated sample, but with the relative ordinate of the line at $H = H_{res} \pm \Delta H/2$. The calculated curve in Fig. 19 was used in the determination of the parameter H_{max} in accordance with Eq. (3.8), and this parameter was related directly to E'. Estimates obtained for the Lorentzian and Gaussian approximations were naturally different. Averaging these estimates gave a value of E' which should have been close to the real value (Table 6).

The additional crystal field in irradiated ruby crystals estimated in this way was much less than the main trigonal field E_i acting on the Cr^{3+} ions. This additional field could be used in estimating the broadening of the optical transition lines in ruby, particularly the R lines, which could then be compared with the experimental data.

The application of an external electric field to ruby splits the R_1 and R_2 lines, and the splitting is proportional to the field intensity [38]. The mechanism of this effect and the nature of the splitting of the ESR lines in an external electric field are related to the existence of two energetically equivalent Cr^{3+} centers located at different lattice sites. The local electric field near the chromium ions can be expressed in the form $\alpha \overline{E}_0$, where $\overline{E}_0$ is the applied electric field, and α is the tensor which is generally governed by the dielectric properties of a crystal. This field should be considered as a correction to the main axial electric field, which gives rise to an additional Stark displacement of the Cr^{3+} ion. Since the two types of Cr^{3+} center are displaced in opposite directions (along the crystal axis) relative to the centers of the anion octahedra, the Stark splitting should be opposite for the levels of these ions. Consequently, we should observe the pseudo-Stark splitting of lines in the optical spectra.

A rigorous calculation of the splitting of the R levels meets with considerable difficulties. An approximate theoretical estimate gives 0.1 cm^{-1} for $E_0 = 10^5$ V/cm. Experimental evidence indicates that the application of an external field of 160 kV/cm to a crystal splits these levels by ~ 1 cm^{-1}.

A qualitative analogy between the splitting of the ESR lines and the splitting of the R lines of ruby allows us to use Eq. (3.5), which is valid for a Lorentzian line profile, in estimating the broadening of one of the R lines emitted by an irradiated crystal under the influence of an additional charge field E'. In this case, the half-width of the R line $\Delta \nu_R$ is given by

$$\Delta \nu_R = \Delta \nu_{R_0} \sqrt{1 + \frac{2k'E'}{\Delta \nu_{R_0}}}, \qquad (3.9)$$

where $\Delta \nu_{R_0}$ is the half-width of a line emitted by an unirradiated sample, and k' is the coefficient of proportionality between the splitting of the R levels and the applied field E_0.

The half-width of the R fluorescence line at 77°K is approximately 0.8 cm^{-1} [38]. A calculation of this half-width for a sample irradiated with a dose of 10^6 rad, carried out using the value of E' obtained from the ESR data, shows that $\Delta \nu_R$ should not exceed the initial value by 0.2 cm^{-1}, i.e., by 25%. At room temperature, the line width for an unirradiated sample is large, and its relative broadening in this range of radiation doses is slight.

It follows from the experimental data on the influence of gamma irradiation on the absorption spectra of ruby in the R-line region [36] that the R lines emitted by samples irradiated with a dose of 10^6 rad are broadened by 0.3-0.4 cm^{-1}. This is in qualitative agreement with the above estimate of the broadening.

§ 3. Paramagnetic Centers in Irradiated Corundum

Bombardment of a ruby crystal with gamma rays alters the charge state not only of the Cr^{3+} ions, but also of other centers which are not associated with chromium. These centers are paramagnetic in irradiated ruby and give rise to an additional ESR spectrum in the form of a single almost isotropic line with g = 2.012. The profile of this line is asymmetric and depends, although weakly, on the orientation of the c crystal axis in the applied magnetic field. This line was exhibited by all our ruby samples with sufficiently low chromium concentrations, and its intensity was approximately the same for all samples that received a given dose. It was difficult to establish the presence of this line in the spectra of ruby containing 0.03% or more chromium because the paramagnetic Cr^{3+} pair centers which appeared in these concentrations produced a strong background ESR spectrum over a wide range, including the region where g ~ 2.0. When a sample was cooled from room temperature to 77°K, the line intensity increased by a factor of approximately 2.5, but the line profile and its angular dependence were not affected. Heating to 250°C destroyed completely the paramagnetic centers.

A line with the same g factor as that found by us for gamma-irradiated ruby was studied by Gamble et al. [25] in the case of corundum irradiated with Co^{60} gamma rays at liquid nitrogen temperature. Gamble et al. established that, for doses exceeding $5 \cdot 10^5$ rad, the intensity of this line tended to saturation. Since the gamma-ray energy was too low to damage the lattice, the results indicated that irradiation caused ionization of structure defects already present in a crystal. However, it was found that the application of strong mechanical loads to a corundum crystal did not enhance significantly the ESR spectrum obtained after irradiation, and, consequently, the observed centers were not due to the lattice deformation.

Gamble et al. also reported that the line profile depended on the orientation of the crystal axis in the magnetic field, and it could be regarded as a superposition of three Gaussian components with different amplitudes and different angular dependences of these amplitudes.

A comparison of the line studied by Gamble et al. with the line emitted by our gamma-irradiated ruby samples indicated that the same paramagnetic centers were responsible for both lines.

The centers responsible for the line with g = 2.012 in our crystals and called arbitrarily the paramagnetic П centers were clearly due to lattice defects. They could be free cation sites which captured holes during irradition. This model explained well the anisotropic ESR spectrum of MgO [39], which was due to centers of structure similar to the II centers in corundum.

In the presence of aluminum vacancies in Al_2O_3, the paramagnetic centers may be formed from an O^- ion of spin 1/2 subjected to a sufficiently strong axial field and adjoining a cation vacancy. In this case, a hole occupies the 2p orbit, and the triply degenerate ground state of O^- is split by the axial field so that the lowest level is an orbital singlet which is doubly spin-degenerate. Since each cation site is surrounded by six nearest anions, there are three different directions of the axis of the additional electric field in a unit cell; this is the field which acts on the paramagnetic centers and is responsible for the appearance of the line components.

Although this model explains the complex ESR spectra of gamma-irradiated metal oxides, a check is needed to ascertain whether it applies to the centers in the corundum lattice because the characteristic time components of such centers cannot be resolved sufficiently clearly in the observed spectra. The question whether cation vacancies are the components of the centers produced by irradiation can be answered by comparing the spectra obtained for crystals with different departures from stoichiometry. Such departures, observed in some compounds with relatively weak atomic binding as a result of prolonged heating in the vapor of the component elements, were extremely difficult to achieve in the case of large Al_2O_3 crystals because of the exceptionally slow diffusion of atoms into these crystals. Therefore, we altered the vacancy concentration by adding excess aluminum to a power sample which had a much larger surface area than a crystal. In a preliminary study it was established that the ESR line of a powder sample due to irradiation was observed under the same conditions as for an irradiated single crystal, and this line had practically the same width as for a crystal because of the weak anisotropy (Fig. 20).

At 1200°C, the temperature used in our experiments, the saturated vapor pressure of aluminum was 1 mm Hg, which was sufficient to maintain a high concentration of the aluminum atoms on the grain surfaces. When a grain was approximated by a sphere of diameter equal to its transverse size, the diffusion of the aluminum atoms could be described sufficiently accurately by [15]

$$C(r,t) = C_0 \frac{r_0}{r}\left[1 - \frac{\pi}{4}\exp\left(-\frac{\pi^2}{r_0^2}Dt\right)\sin \pi \frac{r}{r_0}\right],\tag{3.10}$$

where C_0 is the concentration of the diffusing particles on the surface of the sphere, r_0 is the sphere radius, and t is the heating period. Hence, we concluded that the establishment of a sufficiently homogeneous distribution of the excess aluminum atoms (such that, at the center of a sphere, the concentration of these atoms would be 0.9 of the surface concentration) would require heating for $\sim 10^4$ sec in the case of grains of 10^{-3} cm transverse size with a diffusion coefficient $D = 10^{-10}$ cm^2/sec.

We took into account this estimate and prepared a batch of corundum powder of 50-100 μ grain size by crushing single crystals in an Alundum mortar. In all, three equal-mass samples

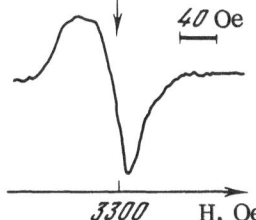

Fig. 20. ESR line with g = 2.012 emitted by gamma-irradiated corundum powder.

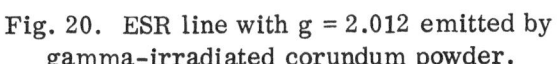

were prepared from this powder, and one was used as a control. The other two samples were mixed separately with aluminum powder, sealed in quartz ampoules, and then heated at 1200°C; one of the samples was heated for 1 h and the other for 2 h. After heating, the samples and control were subjected to gamma irradiation with a dose of $2 \cdot 10^6$ rad. We then recorded the ESR spectra of all three irradiated samples under the same conditions at 77°K. We also recorded a calibration ESR spectrum ($CuSO_4 \cdot 5H_2O$ crystal). The intensity of the g = 2.012 ESR line was considerably lower for the heated samples than for the control. No other lines were observed before and after irradiation. When the intensity of the ESR line of the control sample was taken as unity, it was found that heating for 1 h reduced this intensity to 0.68 and heating for 2 h, to 0.41.

A comparison of the results obtained for different samples indicated that the heating of corundum reduced considerably the concentration of the paramagnetic centers generated by gamma irradiation. This reduction could be explained by the fact that the introduction of excess aluminum into a crystal partly filled the cation vacancies, and this prevented localization of carriers at these vacancies.

An electron compensating the charge of a hole captured by a cation vacancy is localized at some other defect, which may also act as a paramagnetic center. If this defect is an anion vacancy, the ESR spectrum recorded at 77°K should include a structured line with six components due to the interaction between the electron and the Al^{27} nuclei at the neighboring sites [40]. The absence of this line from the observed spectra indicated that corundum crystals contained electron localization centers which were not associated with oxygen vacancies; in particular, they could be monovalent or divalent impurities which did not give rise to the ESR spectra under the conditions in our experiments. Holes were localized at the cation vacancies. Thus, our results indicated that hard or optical irradiation of ruby changed the properties of this material partly because of charging of intrinsic defects such as the cation vacancies.

CHAPTER IV

Mn^{2+} IMPURITY IONS IN THE Na_2ZnGeO_4 LATTICE

One of the interesting new phosphors is sodium zincogermanate Na_2ZnGeO_4. Large single crystals of this compound have been prepared by the hydrothermal synthesis method [41]. Introduction of manganese in amounts of 10^{-4}-10^{-2} g/g ensures bright photoluminescence in the spectral range 500–690 nm when these crystals are excited with ultraviolet radiation.

Manganese-activated sodium zincogermanate exhibits not only photoluminescence, but also triboluminescence, cathodoluminescence, x-ray luminescence [42], dc electroluminescence [43], strong piezoelectric effect, and some phenomena (not observed earlier in crystal phosphors) associated with the unsteady nature of the electroluminescence [44].

Investigations of these crystals have established that they contain paramagnetic centers responsible for a strong ESR spectrum. A general analysis of this spectrum shows that the investigated crystals contain several types of center associated with manganese ions and occupying different positions in the lattice.

Crystals of Na_2ZnGeO_4 belong to the monoclinic system. Their habit is formed by two prisms bounded by the {110} and {011} planes and two pinacoids bounded by the {100} and {001} planes (Fig. 21). Each Zn^{2+} ion is surrounded by four oxygen ions. The ZnO_4 and GeO_4 tetrahedra, between which the Na^+ cations are located, are linked to form a three-dimensional network. The sodium cations can occupy two positions differing in respect of the nearest neighbors. In one of them, an Na^+ ion is surrounded by six oxygen ions, three of which form the

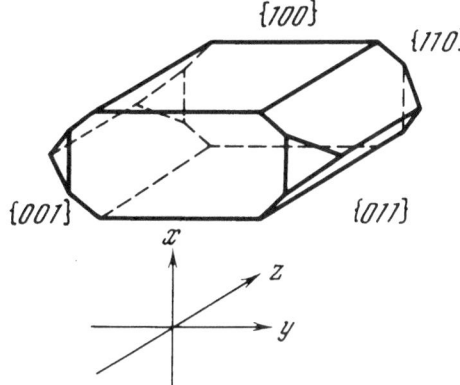

Fig. 21. Natural habit of an Na_2ZnGeO_4 crystal.

first coordination sphere and three, the second sphere. In the other position, an Na^+ ion is surrounded by four oxygen ions, forming an almost regular tetrahedron. The ZnO_4 and GeO_4 tetrahedra are smaller and strongly distorted. A unit cell contains two crystallochemically equivalent complexes which can be transformed into one another by reflection in a plane; therefore, the crystal as a whole has one symmetry plane which is identical with the reflection plane [42].

A detailed study of the luminescence of manganese-activated crystals established that the spectrum depended on the conditions under which the activator was introduced and, particularly, on the compound used to introduce the manganese. When manganese was introduced from MnO_2, the room-temperature luminescence spectrum obtained as a result of excitation with the λ = 365 nm mercury line consisted almost entirely of one green band with a maximum at 2.35 eV. When manganese was introduced from $KMnO_4$, we observed not only green luminescence, but also a wide orange band (the excitation conditions were the same). Since the investigated crystals captured small amounts of many different impurities, the attribution of both luminescence bands to manganese could be made only after special experiments.

An analysis of the spectra by the generalized Alentsev method [21] demonstrated that at 77°K the spectrum consisted of at least four individual bands with maxima at 1.66, 2.00, 2.32, and 2.37 eV. In the spectra of the crystals activated with manganese derived from $KMnO_4$, the intensity of the orange band $h\nu_{max}$ = 2.00 eV was 35-50% of the intensity of the green band $h\nu_{max}$ = 2.37 eV. In the spectra of the crystals activated with manganese from MnO_2, the intensity of the orange band was only a few percent of the band at $h\nu_{max}$ = 2.37 eV. The relative intensities of all four bands observed in the spectra of different crystals activated with manganese from MnO_2 and $KMnO_4$ were compared in [45].

§ 1.   ESR Spectrum of Na_2ZnGeO_4 : Mn

The results of crystallographic investigations of Na_2ZnGeO_4 indicated that the lattice of this compound has at least three symmetric pairs of cation sites where the host atoms can be replaced with metal activator ions. If this activator is manganese, the energetically most favorable isomorphous substitution involves occupation of the Zn^{2+} site because the isovalent Mn^{2+} ion is chemically very stable and has a radius $R_{iMn^{2+}}$ = 0.91 Å close to the radius $R_{iZn^{2+}}$ = 0.93 Å. The Mn^{2+} ions are then located at the centers of two identical oxygen tetrahedra oriented symmetrically with respect to the reflection (symmetry) plane of the crystal. Since these tetrahedra are distorted, the Mn^{2+} ions are subject to a low-symmetry crystal field [46]. Therefore, we may expect the appearance of two independent but identical ESR spectra due to the Mn^{2+} ions at the symmetrically located sites. The presence of a crystal field of axial (or lower) symmetry should cause a slight initial splitting of the spin levels and should give rise to a dependence of the ESR spectrum on the orientation of a crystal in the applied magnetic field.

All the investigated manganese-activated Na_2ZnGeO_4 samples exhibited a strong ESR spectrum both at 77°K and room temperature. The spectrum consisted of a large number of narrow anisotropic lines. There were both strong and weak lines in the spectrum, but the latter were well resolved in certain orientations which ensured a sufficiently large separation between the lines. The total number of the lines (including weak) exceeded 200. A study of the angular dependences indicated that groups of five lines obeyed certain general relationships, and each of these lines had six almost equidistant components of the same intensity. Hence, we concluded that the ESR spectrum was due to the Mn^{2+} ions and that each of the lines in the spectrum was split into six components of the hyperfine structure in accordance with the nuclear spin $I = 5/2$. Some of the lines coincided in pairs when the magnetic field was oriented in the symmetry plane of the crystal or at right angles to this plane.

The two Mn^{2+} complexes of identical structure but with different orientations of the crystal field axes relative to the external magnetic field could give rise to not more than 120 ESR lines. The presence of a much larger number of lines in our spectra indicated that the Mn^{2+} ions also occupied sites other than those of Zn^{2+}.

The complexity of the ESR spectrum and the partial symmetry of the angular dependences, relative to the symmetry plane, could not be explained by the presence of structure defects (such as block structure or irregularities of the ion packing in the lattice) because different samples, including those taken from different batches, exhibited basically identical ESR spectra when the planes were oriented in the same way in the magnetic field.

The centers transforming into one another by symmetry operations and rotation of the axes should naturally have identical luminescence and absorption spectra, since such centers experience a field of the same absolute magnitude which produces the same Stark splitting of the orbital levels. Therefore, these centers should be regarded as identical in respect of the optical properties although their energy states in an external magnetic field are generally different. The nature of complex optical spectra can thus be explained by establishing the presence of centers which differ in respect of interionic distances in the nearest coordination spheres. This requires: 1) determination of the symmetry elements of the crystal; 2) identification of the ESR spectra belonging to different centers; 3) determination of the main constants of the spin Hamiltonian describing the ESR spectra of different centers.

§ 2. Determination of Symmetry Elements of

Na_2ZnGeO_4 Crystals from ESR Spectra of Mn^{2+}

Impurity Ions

Distortion of the oxygen tetrahedron surrounding an Mn^{2+} ion reduces the symmetry of the local crystal field to the orthorhombic form. However, in this case (as also in the case of the axial symmetry of the effective field), the greatest scatter of the ESR lines should be observed when the direction of the external magnetic field coincides with some axis of the tetrahedron. Therefore, when the angle between H and this axis is varied, we should observe angular extrema of the line intensities. For an arbitrary orientation, the position of the fine-structure lines in the ESR spectrum can be described, in the first approximation, by Eq. (1.4). The four lines due to the $M = -5/2 \rightarrow 3/2$, $M = -3/2 \rightarrow -1/2$, $M = 1/2 \rightarrow 3/2$, and $M = 3/2 \rightarrow 5/2$ transitions occurring in the same Mn^{2+} center should have practically identical extrema. The edges of the observed ESR spectrum should be governed by the extremal positions of two of these lines (first and fourth). Thus, all the possible orientations of the tetrahedra and the corresponding MnO_4 complexes, forming a single lattice, determine uniquely the positions of the

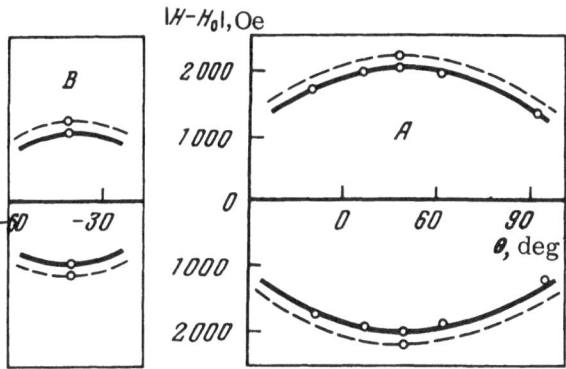

Fig. 22. Angular dependences of the ESR spectra obtained by rotating a crystal about the x axis. The continuous curves represent the $M = {}^3/_2 \to {}^5/_2$ transition lines in the A and B centers, whereas the dashed lines apply to the C centers.

total angular extrema of the ESR lines obtained on rotation of a crystal in an external magnetic field.

The symmetry elements of the lattice can be found by investigating the angular dependences of the ESR spectra obtained by rotating a crystal relative to several axes, each of which is oriented at right-angles to the magnetic field. Since the natural faces form the habit of a crystal with three twofold symmetry axes (x, y, and z in Fig. 21), it is these axes that were first used as the rotation axes in our investigation.

Since the recorded ESR spectrum was a superposition of several overlapping partial spectra, which were difficult to resolve, we determined the angular positions of only those lines which were concentrated in a spectrum near its edges on the low- and high-field sides. When a crystal was rotated relative to the x axis, it was found that the lines had several angular extrema shifted in respect of the angle relative to one another and not repeated within one 180° rotation. We first studied the strongest lines. Figure 22 shows the dependences of the positions of lines of identical intensities on the angle θ between the z axis and the direction H for this type of rotation. These dependences had two extrema, and they indicated that there were two groups of centers oriented in a certain manner for which the projection of the principal crystal field axis on the yz crystal plane was oriented at angles of 49 and −41°. These dependences indicated that none of the planes passing through the rotation axis was a symmetry plane of the crystal lattice. The two groups of centers (A and B) represented the majority of the manganese impurity ions because their lines were the strongest.

When a crystal was rotated about the z axis (also oriented at right-angles to the magnetic field), we identified four lines of identical intensity which was approximately half the intensity in the preceding case. The angular dependence of the ESR spectrum for this rotation axis is plotted in Fig. 23. The positions of the lines had angular extrema which were pairwise symmetric relative to the xz plane. When the orientation was such that H was in this plane, the

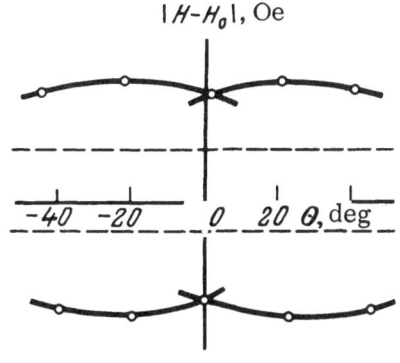

Fig. 23. Angular dependence of the ESR spectrum obtained by rotating a crystal about the z axis. The curves represent the lines due to the $M = {}^3/_2 \to {}^5/_2$ transition in the A and B centers; no measurements were carried out between the horizontal dashed lines.

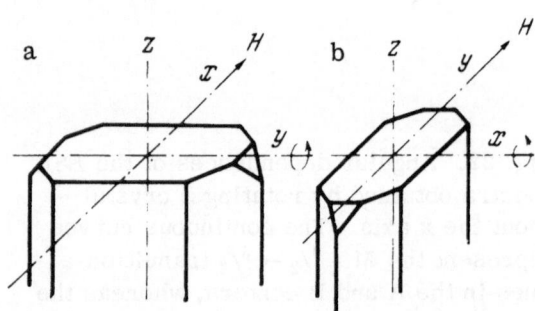

Fig. 24. Orientation of a crystal in determination of the symmetry plane: a) rotation about the y axis; b) rotation about the x axis.

lines coincided in pairs and formed two lines of doubled intensity. Hence, we concluded that each of these groups of centers consisted of two subgroups with equal numbers of centers, and the same angular dependences of the ESR spectra were obtained on rotation of a crystal about the x axis. The symmetry of the angular dependences relative to the xz plane indicated that the crystal lattice had at least one symmetry plane, and this could be under the yz plane or the xz plane perpendicular to the latter because a pairwise coincidence of the lines in the spectrum could be expected in either case.

The position of the symmetry plane of the lattice relative to the faces and morphological axis of the crystal was determined as follows. A crystal was rotated about the z axis until the xz plane was parallel to the field H and the lines coincided in pairs with great accuracy (Fig. 24a). The sample was then rotated through a small angle (1-2°) relative to the y axis which was then directed at right-angles to the field H. It was found that such rotation disturbed the coincidence of the lines and each combined line split into its two equal-intensity components; the displacements of these components relative to the initial position were of different sign. Hence, we concluded that the xz plane was not a symmetry plane of the lattice because when the direction of H was varied in this plane, the angular dependences of the ESR spectra of the equivalent centers differed very considerably.

Similarly, a crystal was rotated about the same z axis until the yz plane was parallel to the vector **H** (Fig. 24b). Then, the sample was rotated through a small angle relative to the x axis which was perpendicular to the vector **H**. The lines in the spectrum were shifted by this procedure in the same direction but, in contrast to the preceding case, they were still coincident. Thus, the operation of rotation in the magnetic field indicated that the only symmetry plane of the host lattice was the yz plane, corresponding to the (100) plane in the adopted coordinate system.

§ 3. Determination of the Orientation of Equivalent Mn^{2+} Centers Relative to Crystal Axes

The angular extrema of the ESR lines of one type of center, observed on rotation of a crystal in a magnetic field relative to some axis, correspond to such an orientation of this center for which the angle between the direction of the axis of the center and the magnetic field has either the smallest or largest value. If the axis of the center is parallel to the magnetic field, the line positions correspond to the principal extremum, i.e., to the greatest scatter of the lines. In this case, the shift of the lines relative to the midpoint of the spectrum is governed uniquely by the parameters of the center, and these are linked to the crystal field. This orientation of a crystal can be used to determine the general paramagnetic properties of the Mn^{2+} centers present as two different groups in the Na_2ZnGeO_4 lattice.

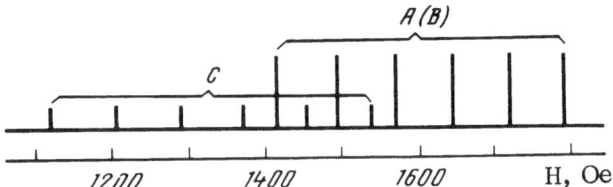

Fig. 25. ESR line of the $M = {}^3/_2 \to {}^5/_2$ transition at the extremum (schematic representation). The components of the lines due to the paramagnetic A and B centers coincide.

The axis of the centers was made parallel to **H** using the angles corresponding to the partial extrema of the lines obtained by rotating a crystal about the x and z axes (Figs. 22 and 23).

Group A centers were oriented along the field H by placing a crystal in a resonator so that its rotation axis was perpendicular to **H** and located in the symmetry plane of the lattice (yz); the angle φ between the z crystal axis and the plane of rotation was 49°. The exact orientation was obtained by ensuring the greatest shifts of the lines from the midpoint of the spectrum by an additional rotation of the crystal in a plane parallel to **H**. In this orientation, the lines furthest from the midpoint of the spectrum were observed in fields H given in Fig. 25. A similar orientation of the group A centers symmetric to that just described produced an identical ESR spectrum.

The same procedure of orientation along the field H was also applied to group B centers. In this case, a crystal in a holder was oriented so that its z axis made an angle $\varphi = -41°$ with the plane of rotation (Fig. 26). When the exact orientation was reached, the line positions were recorded (including the lines furthest from the midpoint of the spectrum). The resonance fields obtained in this way are plotted in Fig. 25. The same values were obtained for the second symmetric orientation of the group B centers along the field H.

The positions of the ESR lines due to the group A and B paramagnetic centers coincided when the orientation was the same in the magnetic field (Fig. 25). Hence, we concluded that, in this case, we were dealing with just one type of Mn^{2+} center, which occupied four positions in the crystal lattice of Na_2ZnGeO_4 differing in respect of the directions of the axes. Therefore, in estimating the concentrations of centers of this type from the ESR spectrum, we should bear in mind that the individual lines observed for an arbitrary orientation of a crystal in the external magnetic field represented one-fourth of the total number of centers.

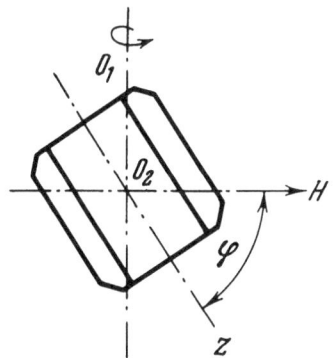

Fig. 26. Orientation of a crystal in the determination of the line intensity extremum; O_1 and O_2 are the axes about which a crystal is rotated.

As mentioned earlier, we observed not only strong lines in the ESR spectrum of Na_2ZnGeO_4 : Mn due to the main type of the Mn^{2+} impurity center, but also (and this was true of all the crystals) a series of weak anisotropic lines whose intensities were approximately an order of magnitude lower. When the orientation in the external magnetic field was such that the scatter (separation) between the strongest lines was greatest, these weak lines were observed clearly at the edges of the overall spectrum (lines denoted by C in Fig. 25). The presence of six almost equidistant components of the same intensity with exactly coincident angular dependences indicated that the weak additional spectrum was also due to the manganese ions. However, it could not be due to the transitions in the main Mn^{2+} centers, and even the forbidden transitions were unlikely. Hence, we concluded that Na_2ZnGeO_4 also contained a different type of center formed by manganese and having different crystallographic properties (group C centers).

The main parameters of the group C centers were determined by recording the angular positions of the extreme lines in the additional ESR spectrum when a crystal was rotated about the x and z axes, exactly as in our study of the distribution of the group A and B centers. It was clear from this dependence (dashed curves in Fig. 22) that the angles corresponding to the extrema of the group C lines coincided with the extrema of the group A and B lines. Determination of the symmetry elements of the crystal lattice for the type C centers, carried out exactly as described above, indicated that the yz plane was a symmetry plane for these centers and that the group C centers also formed four subgroups of equivalent centers differing in respect of the orientation of the elementary axes.

These centers were compared with those of type A and B by recording the low-field part of the ESR spectrum of the C centers in a position corresponding to the principal extremum. The cited analogy between the distributions of the C and A (B) centers in the lattice enabled us to establish this extremal position using the same orientation in the magnetic field for which the principal extremum was obtained for the lines of the A (B) centers. The resonance fields for the C centers (corresponding to the $M = {}^3/_2 \rightarrow {}^5/_2$ transition) obtained for four equivalent orientations of the crystal were much smaller (by 309 ± 2 Oe) than the corresponding resonance fields for the A (B) centers. This indicated a large initial splitting of the spin levels of the C centers. Thus, we concluded that the C centers were not equivalent to the A (B) centers, and they differed somewhat in respect of the crystal field acting on them.

§ 4. Hyperfine-Interaction Constant
of Mn^{2+} Centers

The hyperfine-interaction constant represents the magnetic interaction between electrons in the shell of an ion and its nucleus. To some extent, it reflects the structure of the bonds linking this ion to the neighboring ions in the host lattice. The constant is proportional to the magnetic field generated by electrons at the nuclei. If the bond covalence increases, the electrons are generally located further from the nuclei, and therefore, their interaction becomes weaker. It is shown in [47] that the hyperfine-structure constant A of the $(Mn^{2+})^{55}$ ions decreases linearly with increasing bond covalence. Hence, the Mn^{2+} centers, which occupy positions differing in respect of the interionic distances or in the nature of ions in the immediate environment, should exhibit different hyperfine interactions. Therefore, the value of the constant A may be used to confirm the equivalence or otherwise of the structure of the centers, particularly the A (B) and C centers in the Na_2ZnGeO_4 lattice and this can be done on the basis of their ESR spectra.

The constant A for the centers of both kinds was determined from the hyperfine-structure lines. We used the line positions in the principal extremum when the lines belonging to centers of one subgroup were not overlapped by lines due to other centers. A schematic representation of a part of this spectrum covering only one fine-structure line ($M = ^3/_2 \rightarrow ^5/_2$) of the A (B) and C centers indicated (Fig. 25) that the hyperfine splitting of the lines due to the C centers was somewhat greater than the splitting of the lines of the A (B) centers. The constant A was calculated using the ratio (1.23) describing the positions of the hyperfine-structure lines in the spectrum. The values of this constant were $A_C = 89.0 \pm 2$ Oe and $A_{A(B)} = 84.0 \pm 1$ Oe. These values demonstrated that the ionicity of the Mn^{2+} bonds in the type C centers was somewhat higher, and this confirmed that the C centers were not equivalent to the A (B) centers. If it was assumed that the dependence of the constant A on the bond ionicity [47] was also applicable to Na_2ZnGeO_4, the ionicity of the Mn^{2+} bonds in the C centers was 93% and that in the A (B) centers was 90%.

A comparison of the ESR spectra of Na_2ZnGeO_4 crystals containing activator ions obtained from different compounds ($KMnO_4$, MnO_2) indicated that the ratios of the intensities of the ESR lines of the C and A (B) centers were different for these crystals. In the case of a sample activated with manganese from $KMnO_4$, the ratio of the line peaks, recorded in the integrated form, was 0.14, whereas in the case of a crystal activated from MnO_2, this ratio did not exceed 0.08–0.09. The luminescence spectrum of the former crystal included intense orange bands at $h\nu_{max} = 1.66$ and 2.00 eV. The intensities of these bands in the spectrum of the second crystal were much weaker. A qualitative comparison of the luminescence and ESR spectra of the same samples demonstrated that the long-wavelength bands were, to some extent, due to the type C centers. However, the disagreement between the quantitative data indicated that the origin of the luminescence bands was more complex.

The author is grateful to M. V. Fok and Z. L. Morgenshtern for discussing the results and to N. A. Gorbacheva, T. I. Voznesenskii, and I. P. Kuz'mina for preparation of the samples.

LITERATURE CITED

1. A. Abragam and M. H. L. Pryce, Proc. R. Soc. A, 205:135 (1951).
2. S. A. Al'tshuler and B. M. Kozyrev, Electron Paramagnetic Resonance [in Russian], Nauka, Moscow (1972).
3. H. W. de Wijn and R. F. van Balderen, J. Chem. Phys., 46:1381 (1967).
4. J. Schneider, S. R. Sircar, and A. Räuber, Z. Naturforsch. a, 18:980 (1963).
5. T. V. Anan'eva, K. K. Dubenskii, L. Z. Potvorova, A. I. Ryskin, G. I. Khil'ko, G. K. Chirkin, and L. Ya. Shekun, in: Spectroscopy of Crystals [in Russian], Nauka, Moscow (1970), p. 380.
6. I. V. Shtambur, Thesis for Candidate's Degree [in Russian], Dnepropetrovsk (1969).
7. S. P. Keller, I. L. Gelles, and W. V. Smith, Phys. Rev., 110:850 (1958).
8. H. van den Boom, Rev. Sci. Instrum., 42:524 (1971).
9. K. A. Müller, Helv. Phys. Acta, 28:450 (1955).
10. N. P. Golubeva and M. V. Fok, Zh. Prikl. Spektrosk., 19:851 (1973).
11. B. J. Skinner and P. B. Barton, Am. Mineralogist, 45:612 (1960).
12. L. V. Atroshchenko, F. I. Brintsev, L. A. Sarkisov, and L. A. Sysoev, Izv. Akad. Nauk SSSR, Neorg. Mater., 8:639 (1972).
13. M. Aven and J. A. Porodi, J. Phys. Chem. Solids, 13:56 (1960).
14. S. A. Kostylev and B. N. Sherstyak, Kristallografiya, 8:456 (1963).
15. G. E. Arkhangel'skii, T. I. Voznesenskaya, and M. V. Fok, Kristallografiya, 18:544 (1973).
16. J. Crank, The Mathematics of Diffusion, Clarendon Press, Oxford (1956).
17. P. B. Dorain, Phys. Rev., 120:1190 (1960).
18. R. Lacroix and C. Ryter, Arch. Sci., 9(fasc.spec.):55 (1956).
19. R. S. Title, Phys. Rev., 133:A198 (1964).

20. G. E. Arkhangel'skii, N. A. Gorbacheva, and M. V. Fok, Z . Prikl. Spektrosk., 19:460 (1973).

21. M. V. Fok, Zh. Prikl. Spektrosk., 11:926 (1969).

22. L. F. Vereshchagin, S. V. Starodubtsev, and N. S. Yunusov, Dokl. Akad. Nauk SSSR, 159:300 (1964).

23. C. R. Philbrick, W. R. Davis, and M. K. Moss, Bull. Am. Phys. Soc., 9:499 (1964).

24. T. Maruyama and Y. Matsuda, J. Phys. Soc. Jap., 19:1096 (1964).

25. F. T. Gamble, R. H. Bartram, C. G. Young, O. R. Gilliam, and P. W. Levy, Phys. Rev., 134:A589 (1964).

26. N. V. Karlov and A. A. Manenkov, Quantum Amplifiers [in Russian], VINITI, Moscow (1966).

27. R. H. Hoskins and B. H. Soffer, Phys. Rev. 133:A490 (1964).

28. D. R. Mason and J. S. Thorp, Proc. Phys. Soc. Lond., 87:49 (1966).

29. K. Morigaki, J. Phys. Soc. Jap., 19:187 (1964).

30. G. E. Arkhangel'skii, Z. L. Morgenshtern, and V. B. Neustruev, Phys. Status Solidi, 22:289 (1967).

31. G. E. Arkhangel'skii, Z. L. Morgenshtern, and V. B. Neustruev, in: Spectroscopy of Crystals [in Russian], Nauka, Moscow (1970), p. 273.

32. Z. L. Morgenshtern and V. B. Neustruev, ZhETF Pis'ma Red., 2:507 (1965).

33. V. B. Neustruev, Thesis for Candidate's Degree [in Russian], Moscow (1972).

34. A. A. Manenkov and V. B. Fedorov, Zh. Eksp. Teor. Fiz., 38:1042 (1960).

35. N. S. Yunusov, Thesis for Candidate's Degree [in Russian], Tashkent (1966).

36. G. E. Arkhangel'skii, Z. L. Morgenshtern, and V. B. Neustruev, Phys. Status Solidi, 36:451 (1969).

37. A. I. Ritus and A. A. Manenkov, Fiz. Tverd. Tela, 5:3590 (1963).

38. W. Kaiser, S. Sugano, and D. L. Wood, Phys. Rev. Lett., 6:605 (1961).

39. J. E. Wertz, J. W. Orton, and P. Auzins, Discuss. Faraday Soc., No. 31, 140 (1961).

40. S. Y. La, R. H. Bartram, and R. T. Cox, J. Phys. Chem. Solids, 34:1079 (1973).

41. I. P. Kuz'mina, O. K. Mel'nikov, and B. N. Litvin, in: Hydrothermal Synthesis of Crystals (ed. by A. N. Lobachev), Consultants Bureau, New York (1971), p. 99.

42. É. A. Kuz'min, V. V. Il'yukhin, and N. V. Belov, Kristallografiya, 13:976 (1968).

43. K. A. Verkhovskaya, I. P. Kuz'mina, A. N. Lobachev, and V. M. Fridkin, Fiz. Tverd. Tela, 10:1906 (1968).

44. M. V. Fok and E. Yu. L'vova, ZhETF Pis'ma Red., 13:346 (1971); Tr. Fiz. Inst. Akad. Nauk SSSR, 68:95 (1973).

45. G. E. Arkhangel'skii, E. Yu. L'vova, and M. V. Fok, Zh. Prikl. Spektrosk., 14:97 (1971).

46. I. P. Kuz'mina, A. N. Lobachev, V. M. Vinokurov, N. I. Nizamutdinov, and L. A. Volkova, Kristallografiya, 18:180 (1973).

47. O. N. Matumura, J. Phys. Soc. Jap., 14:108 (1959).

APPLICATION OF THE POLARIZATION DIAGRAM METHOD TO STUDIES OF UNIAXIAL CRYSTALS

E. E. Bukke, N. N. Grigor'ev, and M. V. Fok

The method of luminescence polarization diagrams was extended to birefringent (uniaxial) crystals. Formulas were derived for polarization diagrams applicable to different orientations of radiators. A study was made of the influence of absorption (including dichroism) and of various depolarizing factors. The general form of a dipole radiator was considered on the assumption that it consisted of a linear electric dipole and and a rotator sharing the axis with the dipole. This model was checked against the results of an investigation of the blue luminescence centers in zinc sulfide. Differences were found between the statistical weights of radiators oriented at different angles with respect to the stacking fault plane.

Introduction

Luminescence centers in crystal phosphors may be single atoms or ions or they may be pairs which are so close to one another that they absorb and emit light as a single quantum-mechanical system. Pairs of this kind are frequently formed from donors and acceptors because in the ionized state they have opposite charges relative to the lattice, and during growth of a crystal they drift toward one another because of the mutual attraction. A pair with the shortest internal distance usually occupies neighboring cation and anion lattice sites, but in other pairs the components may be separated by longer distances: For example, components may be located in adjacent coordination spheres or they may occupy two neighboring like lattice sites, i.e., two cation or two anion sites.

In all these cases the luminescence centers have a definite orientation represented by the line joining the nuclei of the atoms forming a pair. For any one direction in a crystal lattice we can find one or more equivalent (in the crystallographic sense) directions which are linked by the lattice symmetry transformations. Therefore, luminescence centers with identical properties may be oriented in different ways in the crystal lattice. This also applies to the case when one or both atoms (ions) of a pair are located not at lattice sites, but at interstices, since interstitial positions in the lattice are distributed as regularly as the sites. Moreover, one or both components of a pair may be not an atom, but a vacancy (vacant site). Moreover, if we bear in mind that the luminescence centers in crystal phosphors may be formed by many impurities, it becomes clear that a very wide assortment of centers may be present in a given crystal lattice. This is why there are usually several contradictory views about the origin of any one luminescence band.

A comprehensive investigation, which yields several parameters of luminescence centers, is needed in order to determine which of the possible models of a center corresponds to reality. In particular, it is very desirable to know the orientation of luminescence centers in the crystal lattice. This makes it possible to eliminate some a priori possible models because a definite orientation (or a lack of orientation if the center is formed by a single atom) corresponds to each model. In this elimination process we can use the polarization diagram method developed many years ago by P. P. Feofilov [1] and used successfully since in studies of cubic crystals.

The method can be described briefly as follows. A plane-parallel plate, oriented in a definite way relative to the crystallographic axes, is cut from a larger crystal. Plane-polarized light, incident normally on the plate surface, excites luminescence which is observed from the opposite side along the normal. An analyzer is placed in the path of the luminescence, and the intensity of the transmitted luminescence is measured for two positions of the analyzer, in one of which the electric vector of the transmitted luminescence is parallel to the electric vector of the exciting light incident on the opposite side of the plate ($I_\parallel$), whereas in the other position these two vectors are mutually perpendicular ($I_\perp$). These measurements are used to calculate the polarization ratio

$$P = \frac{I_\parallel - I_\perp}{I_\parallel + I_\perp}. \tag{1}$$

The plane of polarization of exciting light is then rotated through an angle φ relative to the crystal, and the measurements are repeated. The dependence $P(\varphi)$ is obtained in this way. Feofilov found that this dependence is exhibited even by isotropic cubic crystals, and it is different for different elementary radiators (electric or magnetic dipoles, quadrupoles, etc.) and for different orientations of these radiators in a crystal.

The value of P depends on the angle φ because the absorption of the exciting light by the luminescence centers depends on the orientation of these centers relative to the electric vector of this light. This gives rise to some anisotropy of the excitation process, which is enhanced by the fact that the luminescence emitted by the centers under consideration also depends on their orientation. Consequently, luminescence is partly polarized, and the degree of its polarization, as well as the predominant direction of the electric vector, depend on the orientation of the plane of oscillation of the electric vector of the incident light. The former effect gives rise to different values of $I_\parallel$ and $I_\perp$ and the latter to the dependence of this difference on the angle φ.

It is worth noting that the polarization ratio is not always equal to the degree of polarization of luminescence. For example, let us assume there is only one possible orientation of absorption and luminescence dipoles in a crystal. Then, luminescence is always plane-polarized, and only its intensity depends on the angle φ. Therefore, if P were the degree of polarization of luminescence, it should then be independent of φ and equal to unity. However, it is clear from Eq. (1) that P = 1 only when $I_\perp = 0$ and $I_\parallel \neq 0$. In our case this occurs only when the analyzer transmits light whose electric vector is parallel to the corresponding vector of the emitted luminescence. If the analyzer is rotated through 45° (strictly speaking, we should rotate also the plane of polarization of the exciting light although in this case such rotation does not affect the luminescence polarization), it is found that $I_\parallel = I_\perp$ and, consequently, P = 0. Although it is not logical to regard as the degree of polarization a quantity which vanishes for completely polarized light, the ratio P is called by some authors the degree of polarization. We shall use the term "polarization ratio."

It is found experimentally that the angular dependence of the polarization ratio for cubic crystals is indeed different in different cases but its absolute value is considerably less than

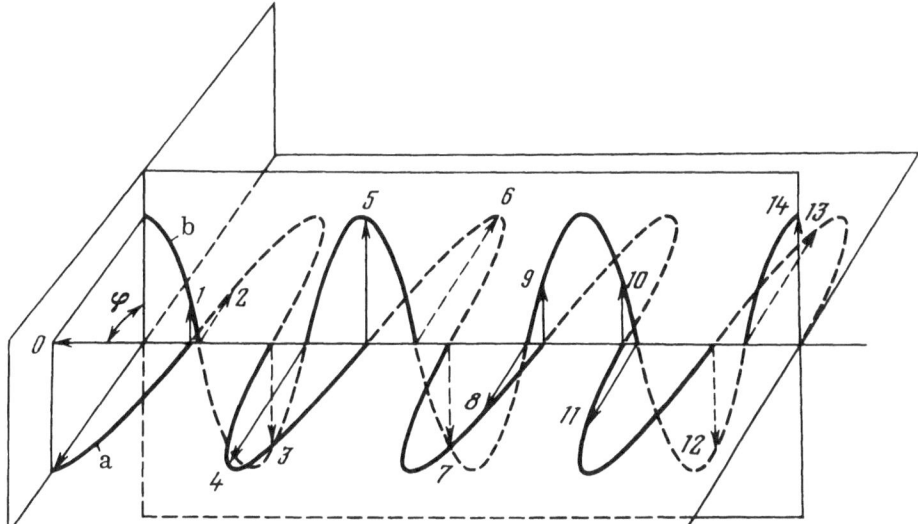

Fig. 1. Electric vectors of light at different points in a crystal: a) position of the end of the electric vector of the ordinary ray; b) corresponding
position for the extraordinary ray; φ is the angle between the optic axis
and the direction of the electric vector of the light incident on a crystal;
0–14 denote the positions of the net electric vector (the points are num-
bered away from the entry surface).

the value predicted theoretically. This disagreement is due to depolarization. As mentioned
earlier, all these results have been obtained for cubic crystals. Our task is to extend the
polarization diagram method to birefringent (uniaxial) crystals.

§1. Allowance for Birefringence

We shall consider a plane-parallel plate cut from a uniaxial crystal so that its optic axis
lies in the plane of the plate. We shall assume that this crystal contains oscillators with a
fixed direction of the axis, capable of absorbing and emitting light. We shall assume that the
axes of the absorbing and emitting oscillators are identical but their frequencies are different.
The first of these assumptions reflects the observation that the symmetry of the distribution
of atoms or ions surrounding a luminescence center is, in the first approximation, independent of
whether this center is excited or not. The second assumption simply allows for the well-known
Stokes shift. It allows us to ignore interference between the exciting light and luminescence.

Plane-polarized light is converted by a birefringent crystal into elliptically polarized
light, and the parameters of the ellipse described by the end of its electric vector vary during
propagation in a crystal (Fig. 1). Therefore, oscillators located at different distances from
the entry surface are excited with different probabilities even when they are oriented exactly
parallel to one another. When these oscillators are excited, they are capable of emitting
plane-polarized light which is also converted into elliptically polarized light,* and the parameters of the ellipse vary with the distance from the emitting oscillator.

---

*We are ignoring here the influence of birefringence on the probability of emission of light
along a given direction and on the polarization of such light. The necessary corrections are
small because usually the difference between the refractive indices n_0 and n_e does not exceed
1%.

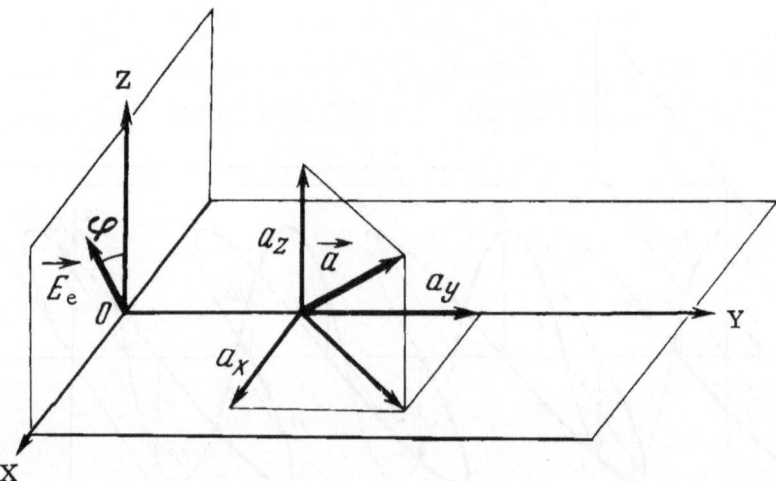

Fig. 2. Coordinate system in a crystal and orientation of an oscillator. Here, OZ is the direction of the optic axis of the crystal; OY is the direction of observation; a is a unit vector representing the direction of the oscillator; a_x, a_y, and a_z are the projections of this vector onto the coordinate axes; E_e is the electric vector of the exciting light.

Luminescence emitted by any one elementary oscillator retains its complete elliptic polarization on leaving the crystal. However, light from many identically oriented oscillators is only partly polarized because there is no coherence between the oscillators, and the parameters of the ellipse described by the electric vector of the luminescence are different for each oscillator because the distances traveled by luminescence inside the crystal are different.

We shall now describe this situation mathematically. We shall assume that all the oscillator parameters are known. Therefore, we can ignore the details of the structure of the luminescence centers and employ the quasiclassical approximation.

We shall select the coordinate system so that the X and Z axes lie in the plane of the plate whereas the Z axis is parallel to the optic axis. The Y axis is assumed to be directed along the light beam (Fig. 2). We shall use φ to denote the angle between the Z axis and the electric vector of the exciting light. Then, the electric field of the exciting light $\mathbf{E}_e$ inside the plate at a distance y from edge has the following components:

$$E_{ex} \propto \sin \varphi \cos \omega t, \qquad E_{ey} = 0, \qquad E_{ez} \propto \cos \varphi \cos (\omega t + \delta_e y), \tag{2}$$

where $\delta_e y$ is the path difference between the ordinary and extraordinary rays of the exciting light, acquired by traveling a distance y.

Let us assume that at a distance y from the plate surface there is a linear oscillator (dipole), whose orientation is described by a unit vector $\mathbf{a} = \{a_x, a_y, a_z\}$. The oscillation energy W of this oscillator is proportional to the mean-square value of the electric field component parallel to $\mathbf{a}$:

$$W_{osc} \propto \overline{(\mathbf{E}_e \mathbf{a})^2} \propto \overline{[a_x \sin \varphi \cos \omega_e t + a_z \cos \varphi \cos (\omega_e t + \delta_e y)]^2}. \tag{3}$$

Squaring and averaging yields

$$W_{osc} \propto a_x^2 \sin^2 \varphi + a_z^2 \cos^2 \varphi + 2a_x a_z \sin \varphi \cos \varphi \cos \delta_e y. \tag{4}$$

The oscillation amplitude of this oscillator is proportional to $\sqrt{W_{osc}}$. The electric vector of the luminescence E_l, emitted along the Y axis, is proportional to this oscillation amplitude and lies in a plane formed by the Y axis and the vector a. The luminescence emerging from the plate has the following components:

$$E_{lx} \sim \sqrt{W_{osc}}\, a_x \cos \omega t, \quad E_{ly} = 0, \quad E_{lz} \sim \sqrt{W_{osc}}\, a_z \cos[\omega t + \delta_l(d - y)], \qquad (5)$$

where d is the thickness of the plate, and the term $\delta_1 (d - y)$ is equal to the phase difference between E_{lx} and E_{lz} acquired in the distance traveled between the oscillator and the plate boundary.

An analyzer which passes light with the electric vector oriented at an angle ψ with respect to the Z axis transmits a flux whose intensity is proportional to

$$i_1 \propto (\overline{E_{lx}\sin\psi + E_{lz}\cos\psi})^2 \propto W_{osc}\,[a_x^2\sin^2\psi + a_z^2\cos^2\psi + 2a_x a_z \sin\psi\cos\psi\cos\delta_1(d-y)], \qquad (6)$$

where W_{osc} is defined by Eq. (4).

If we can ignore the attenuation of the exciting light and of the emerging luminescence as a result of absorption in the plate, a quantity proportional to the intensity of the luminescence generated by all the oscillators of a given orientation can be found by integrating $i_1(y)$ with respect to dy. We shall add the intensities and not the amplitudes because the oscillators emit light spontaneously, and their radiation is noncoherent. Integration and division by d, for the purpose of changing to dimensionless variables, yields

$$\frac{1}{d}\int_0^d i_1(y)\,dy \propto I_1 = (a_x^2\sin^2\varphi + a_z^2\cos^2\varphi)(a_x^2\sin^2\psi + a_z^2\cos^2\psi) +$$
$$+ 2a_x a_z (a_x^2\sin^2\varphi + a_z^2\cos^2\varphi)\sin\psi\cos\psi\,\frac{\sin\delta d}{\delta d} +$$
$$+ 2a_x a_z (a_x^2\sin^2\psi + a_z^2\cos^2\psi)\sin\varphi\cos\varphi\,\frac{\sin\delta d}{\delta d} +$$
$$+ 2a_x^2 a_z^2 \sin\varphi\cos\varphi\sin\psi\cos\psi\left(\frac{\sin\delta d}{\delta d} + \cos\delta d\right). \qquad (7)$$

An additional simplification is made in the derivation of this formula: It is assumed that $\delta_e = \delta_l$, i.e., that the dispersion of the birefringence can be ignored in the spectral region under consideration. Therefore, δ has no index in Eq. (7). If necessary, we can derive also the formula for I_1 when $\delta_e \neq \delta_l$, but we shall not do this because the results become unwieldy.

Substituting $\psi = \varphi$ or $\psi = \varphi + \pi/2$ in Eq. (7), we can calculate the functions $I_{1\parallel}(\varphi)$ and $I_{1\perp}(\varphi)$ and then find their sum and difference. Elementary but fairly lengthy calculations yield

$$I_{1\parallel}(\varphi) - I_{1\perp}(\varphi) = \frac{1}{2}(a_z^4 - a_x^4)\cos 2\varphi + \frac{1}{2}(a_z^2 - a_x^2)^2\cos^2 2\varphi +$$
$$+ a_x a_z (a_x^2 + a_z^2)\sin 2\varphi\,\frac{\sin\delta d}{\delta d} + a_x a_z (a_z^2 - a_x^2)\sin 4\varphi\,\frac{\sin\delta d}{\delta d} + a_x^2 a_z^2\sin^2 2\varphi\left(\frac{\sin\delta d}{\delta d} + \cos\delta d\right), \qquad (8)$$

$$I_{1\parallel}(\varphi) + I_{1\perp}(\varphi) = \frac{1}{2}(a_x^2 + a_z^2)^2 + \frac{1}{2}(a_z^4 - a_x^4)\cos 2\varphi + a_x a_z (a_x^2 + a_z^2)\sin 2\varphi\,\frac{\sin\delta d}{\delta d}. \qquad (9)$$

If there are j possible different orientations of equivalent oscillators in a crystal, we have to derive expressions such as Eqs. (8) and (9) for each orientation and then sum them with respect to j. Dividing the sums we then obtain the following formula for the polarization ratio:

$$P(\varphi) = \sum_j [I_{j\parallel}(\varphi) - I_{j\perp}(\varphi)] : \sum_j [I_{j\parallel}(\varphi) + I_{j\perp}(\varphi)]. \qquad (10)$$

We shall now consider Eqs. (8)-(10) in greater detail. We can see that the polarization ratio is independent of the refractive indices of the ordinary and extraordinary rays considered separately, and it varies only with the phase shift between these rays, which is acquired over the whole thickness of the plate. This can be used in experimental studies by selecting the crystal thickness so that δd has the desired value.

It is clear from these formulas that $P(0)$ and $P(90°)$ are independent of the birefringence because at these points we have $\sin 2\varphi = \sin 4\varphi = 0$, and only the coefficients in these trigonometric functions depend on δd. This important property of the polarization diagrams makes it much easier to compare the theory with experiment because the birefringence is not always known.

The terms depending on δd play an important role in the range $\varphi = 45$ and $135°$ because at these points we have $\cos 2\varphi = 0$, and, consequently, all the terms independent of δd vanish. Since δd occurs in the remaining terms in a sine or cosine, it follows that the value of P at these points depends nonmonotonically on δd, but the variation is not completely periodic. (When δd increases, the coefficients in front on $\sin 2\varphi$ and $\sin 4\varphi$ tend to zero, but the coefficient in front of $\sin^2 2\varphi$ does not show this tendency. Therefore, the shape of the polarization diagram depends on δd even for large values of δd.)

Near the points where $\varphi = 45$ and $135°$ we can clearly see the asymmetry of the polarization diagram relative to $\varphi = 90°$, because at these points $\sin 2\varphi$ is $+1$ and -1, respectively, and the majority of the "symmetric" terms vanish.   However, if oscillators are distributed in pairs symmetrically with respect to a plane perpendicular to the optic axis of the crystal (they can also be symmetric relative to the ZY plane), this asymmetry disappears. In this case both coordinates a_x and a_z of one of the oscillators in a pair are equal (in the absolute sense) to the corresponding coordinates of the other oscillator. One (and only one) of these coordinates differs in sign between the two oscillators in a pair. Therefore, if we sum over all the oscillators, the terms containing the products $a_x a_z$ cancel out, and these are the terms which are asymmetric relative to the point $\varphi = 90°$. The same result is naturally obtained also when the oscillators are distributed so that $a_x = 0$ or $a_z = 0$. This case can be regarded as the limit when two oscillators located symmetrically relative to the XY (or ZY plane) have merged into one.

This feature of the polarization diagrams allows us to determine experimentally how symmetrically the oscillators are oriented relative to the XY plane even without determining the actual oscillator orientation. If the polarization diagram is asymmetric, in spite of the fact that all the necessary experimental conditions are fulfilled (i.e., all possible external causes of asymmetry are removed: the plate is not wedge-shaped, it is not inclined with respect to the optic axis, etc.), we can say that oscillators are distributed asymmetrically. However, if the diagram is symmetric, we have to ensure additionally that the ratio $(\sin \delta d)/\delta d$ is not small and that the corresponding terms play a significant role.

The value of δd can be found by crystal-optics methods. It is easiest to use the apparatus in which the polarization diagram is measured and to illuminate the crystal with light of the same wavelength as the luminescence. This light passes through the crystal without absorption, but its polarization is converted from planar to elliptic. Next, we should measure the polarization diagram in the same way as for luminescence. We can easily show* that in this case we have

$$P(\varphi) = \cos^2 2\varphi + \sin^2 2\varphi \cos \delta d. \tag{11}$$

---

* This can be done simply bearing in mind that light emerging from a crystal has complete elliptic polarization with the components

$$E_x \propto \sin \varphi \cos \omega t, \qquad E_z \propto \cos \varphi \cos (\omega t + \delta d),$$

and then use these components in Eq. (6) instead of E_{lx} and E_{lz} in the calculation of I_i, which is then found to be I_1 because there is no need for integration. Knowing I_1, we can calculate P without any difficulty.

This function has minima at $\varphi = 45$ and $135°$ exactly, and at these points we find that $P = \cos \delta d$. Hence, we can also find the value of δd, but the answer is many-valued. We can select the value of δd corresponding to reality by reducing somewhat the thickness of a crystal (for example, by grinding) and then finding $\cos \delta d$ and a new series of possible values of δd. Comparing it with the first series, we can find a pair of values of δd whose ratio is equal to the ratio of the corresponding thicknesses. This selection is completely unambiguous if as a result of grinding the value of δd does not change by more than 2-3 rad.

It also follows from Eq. (11) that in this case $P(0) = P(90°) = 1$. Hence, we can find experimentally how perfect the investigated crystal is and whether there is any depolarization due to, for example, the scattering of light.

Thus, if the absorption of exciting light and of the luminescence is sufficiently weak, the birefringence is not an obstacle to the application of the polarization diagram method in studies of the orientation of elementary radiators in crystals although the shape of the polarization diagrams depends strongly on the birefringence. The birefringence does not destroy the polarization pattern because the combination of the shapes and orientations of the ellipse, described by the end of the electric vector of the exciting light (and of the luminescence) at different points in a crystal, does not exhaust all the possible sets of the shapes and orientations of an ellipse. Therefore, the dependence of the degree of excitation of elementary oscillators on their orientation is retained on the average throughout the crystal.

§ 2. Allowance for Absorption of
Exciting Light and Luminescence

Strictly speaking, the reasoning in the preceding section is self-contradictory. After all, luminescence cannot be excited without the absorption of the exciting light. Therefore, we must determine the implications of ignoring the absorption of exciting light and find the conditions under which such absorption can be neglected.

In the preceding section we have ignored not the absorption itself, but the attenuation of the exciting light and of luminescence due to absorption. Clearly, this attenuation can be made as small as we please if the crystal plate is made sufficiently thin. However, this also weakens the luminescence intensity. Therefore, it is not always possible to establish such experimental conditions that the excitation is sufficiently homogeneous throughout the sample. Moreover, we may encounter dichroism of the exciting light and of luminescence, as a result of which an ordinary ray is attenuated more strongly than an extraordinary ray or conversely; this can also give rise to a preferential direction of the electric vector of light. A strong dichroism of the emitted luminescence is particularly undesirable because then the luminescence polarization is independent of the polarization of the exciting light and of the orientation of the elementary radiators. Fortunately, luminescence is usually absorbed less strongly than the exciting light. Therefore, in most cases it is possible to select the plate thickness in such a way that the absorption of the luminescence is negligible or can be allowed for by a correction.

We shall allow for the attenuation of the exciting light by introducing certain factors in the expressions for E_{ex} and E_{ez} in Eq. (2) so that these expressions become

$$\left.\begin{aligned}
E_{ex} &\propto \exp\left(-\frac{K_{ex}y}{2}\right) \sin \varphi \cos \omega t, \\
E_{ez} &\propto \exp\left(-\frac{K_{ez}y}{2}\right) \cos \varphi \cos (\omega t + \delta_e y)
\end{aligned}\right\} \tag{12}$$

where K_{ex} and K_{ez} are the coefficents of absorption of the exciting light with the electric vector parallel to the X and Z axes, respectively. These expressions also allow for the dichroism because we may have $K_{ex} \neq K_{ez}$. The coefficient 1/2 is introduced because we are discussing a reduction in the amplitude of the electric field and not of the intensity of light, which is known to be proportional to the square of the amplitude.

Consequently, the expression (4) for the oscillation energy of one oscillator becomes

$$W \propto \exp\left(-K_{ex}y\right)a_x^2 \sin^2\varphi + \exp\left(-K_{ez}y\right)a_z^2\cos^2\varphi +$$

$$+ 2\exp\left(-\frac{K_{ex}+K_{ez}}{2}y\right)a_x a_z \sin\varphi \cos\varphi \cos\delta_e y. \tag{13}$$

Similar factors have to be introduced into the expressions for E_{lx} and E_{lz} in Eq. (5):

$$\left.\begin{array}{l} E_{lx} \propto \sqrt{W}\exp\left[-\dfrac{K_{lx}(d-y)}{2}\right]a_x\cos\omega t, \\[2mm] E_{lz} \propto \sqrt{W}\exp\left[-\dfrac{K_{lz}(d-y)}{2}\right]a_z\cos\left[\omega t + \delta_l(d-y)\right], \end{array}\right\} \tag{14}$$

where W is calculated using Eq. (13).

Such calculations are time-consuming but they can be carried out in the same way as in the absence of absorption, and we can obtain expressions for $I_{1\|}(\varphi) - I_{1\perp}(\varphi)$ and $I_{1\|}(\varphi) + I_{1\perp}(\varphi)$, which generalize Eqs. (8) and (9) to the case of dichroism which is different for the exciting light and for luminescence (and also to the case when $\delta_e \neq \delta_l$). These calculations present no fundamental difficulties but they are cumbersome and the expressions obtained are unwieldy. Therefore, we shall not give them in their general form but consider only the most important practical case when there is no absorption of the luminescence, i.e., when K_{lx} and K_{lz} vanish.

Moreover, as before, we shall assume that $\delta_e = \delta_l$. In this case, we find that

$$I_{1\|}(\varphi) - I_{1\perp}(\varphi) = (a_z^2 - a_x^2)a_x^2\cos 2\varphi(1 - \cos 2\varphi)\frac{1 - \exp(-K_{ex}d)}{2K_{ex}d} +$$

$$+ (a_z^2 - a_x^2)a_z^2\cos 2\varphi(1 + \cos 2\varphi)\frac{1 - \exp(-K_{ez}d)}{2K_{ez}d} +$$

$$+ a_x^3 a_z\sin 2\varphi(1 - \cos 2\varphi)\frac{\delta d\sin\delta d + K_{ex}d\left[\cos\delta d - \exp(-K_{ex}d)\right]}{\delta^2 d^2 + K_{ex}^2 d^2} +$$

$$+ a_x a_z^3\sin 2\varphi(1 + \cos 2\varphi)\frac{\delta d\sin\delta d + K_{ez}d\left[\cos\delta d - \exp(-K_{ez}d)\right]}{\delta^2 d^2 + K_{ez}^2 d^2} +$$

$$+ (a_z^2 - a_x^2)a_x a_z\sin 2\varphi\cos 2\varphi\left\{\frac{\delta d\sin\delta d\exp\left(-\dfrac{K_{ex}+K_{ez}}{2}d\right)}{\delta^2 d^2 + \dfrac{1}{4}(K_{ex}+K_{ez})^2 d^2} + \right.$$

$$\left. + \frac{\dfrac{K_{ex}+K_{ez}}{2}d\left[1 + \exp\left(-\dfrac{K_{ex}+K_{ez}}{2}d\right)\cos\delta d\right]}{\delta^2 d^2 + \dfrac{1}{4}(K_{ex}+K_{ez})^2 d^2}\right\} +$$

$$+ 2a_x^2 a_z^2\sin^2 2\varphi\left\{\frac{\cos\delta d\left[1 - \exp\left(-\dfrac{K_{ex}+K_{ez}}{2}d\right)\right]}{(K_{ex}+K_{ez})d} + \right.$$

$$+ \frac{\delta d\sin\delta d\left[1 + \exp\left(-\dfrac{K_{ex}+K_{ez}}{2}d\right)\right]}{4\delta^2 d^2 + \dfrac{1}{4}(K_{ex}+K_{ez})^2 d^2} +$$

$$\left. + \frac{\dfrac{1}{4}(K_{ex}+K_{ez})d\left[1 - \exp\left(-\dfrac{K_{ex}+K_{ez}}{2}d\right)\right]}{4\delta^2 d^2 + \dfrac{1}{4}(K_{ex}+K_{ez})^2 d^2}\right\}, \tag{15}$$

$$I_{1\parallel}(\varphi) + I_{1\perp}(\varphi) = (a_x^2 + a_z^2)\,a_x^2\,(1 - \cos 2\varphi)\,\frac{1 - \exp(-K_{ex}d)}{2K_{ex}d} +$$

$$+\,(a_x^2 + a_z^2)\,a_z^2\,(1 + \cos 2\varphi)\,\frac{1 - \exp(-K_{ez}d)}{2K_{ez}d} +$$

$$+\,a_x a_z\,(a_x^2 + a_z^2)\sin 2\varphi \left\{ \frac{\delta d \sin \delta d \exp\left(-\dfrac{K_{ex} + K_{ez}}{2}\,d\right)}{\delta^2 d^2 + \dfrac{1}{4}(K_{ex} + K_{ez})^2 d^2} + \right.$$

$$\left. +\,\frac{\dfrac{1}{2}(K_{ex} + K_{ez})\,d\left[1 - \exp\left(-\dfrac{K_{ex} + K_{ez}}{2}\,d\right)\cos \delta d\right]}{\delta^2 d^2 + \dfrac{1}{4}(K_{ex} + K_{ez})^2 d^2} \right\} \qquad (16)$$

These formulas differ from Eqs. (8) and (9) also because the terms are grouped in different ways so that the influence of absorption can be followed more easily. The dependences of the coefficients in front of the trigonometric functions of the angle 2φ on a_x and on a_z remain as before but all these coefficients include two new parameters $K_{ex}d$ and $K_{ez}d$. They can be determined in independent experiments and substituted into the above formulas, so that the function $P(\varphi)$ can still be calculated for any orientation of the oscillators. Morevoer, some features of the polarization diagrams, discussed above, are still retained: $P(0)$ and $P(90°)$ are independent of the birefringence; the polarization diagrams are asymmetric near the points $\varphi = 45$ and $135°$, and the influence of the birefringence is strongest near these points. However, the shape of the polarization diagrams may vary considerably with rising absorption. Only the values of $P(0)$ and $P(90°)$ remain constant because at these points there is just one term in the numerator and denominator of the expression for P, and the factors depending on the absorption cancel out. The physical meaning is clear: When the electric vector of light is parallel or perpendicular to the optic axis of a crystal, a beam does not split into two, and the system behaves as if there were no birefringence or dichroism. On the other hand, at $\varphi = 45$ and $135°$ the intensities of the ordinary and extraordinary rays are the same at the entry into a crystal. Their ratio may change only because of dichroism, and, therefore, the influence of dichroism at these points is greatest. This influence is particularly strong for $\delta d = (2n + 1)\pi$. If the absorption is ignored, the asymmetric terms vanish but in the presence of absorption they may not be small compared with the other terms even if the absorption is weak. For example, if $\delta d = \pi$, $K_{ex}d = K_{ez}d \leq 0.3$, and $a_x^2 = a_z^2$, we find that to within several percent

$$P(\varphi) = -\frac{2Kd}{\pi^2}\sin 2\varphi + \frac{1}{4}\sin^2 2\varphi. \qquad (17)$$

If $Kd = 0.3$, the coefficient in front of the first term is only four times smaller than the coefficient in front of the second term. This means that $P(45°)$ and $P(135°)$ differ by a factor of almost 2 (whereas for $Kd = 0$ they are equal).

All this shows that in applying the polarization diagram method we have to know not only δd, but also the values of $K_{ex}d$ and $K_{ez}d$, even if they are several times smaller than unity. They can be determined, for example, from the difference between the attenuations of the exciting light in a crystal plate before and after reduction of its thickness. Measurements for two thicknesses are essential also for the determination of δd, so that this does not cause any additional difficulties.

If a crystal is so thick that the exciting light is absorbed almost completely in the investigated plate, Eqs. (15) and (16) simplify considerably because for $Kd \gg 1$ all the exponential terms can be ignored. However, even in the absence of dichroism the polarization diagram continues to depend on Kd. Since it is difficult to find Kd for such thick crystals, it is preferable to reduce the thickness of the crystal plate especially as this has little effect on the total luminescence intensity.

The exception to this rule is the case when the dichroism is absent, $Kd \gg 1$, and δd is such that

$$\frac{\delta d}{Kd} \ll \cot \delta d \tag{18}$$

and

$$\frac{\delta d}{Kd} \ll \frac{1}{4}(\cot \delta d + \csc \delta d). \tag{19}$$

In this case the polarization diagram is completely independent of the absorption:

$$I_{1\parallel}(\varphi) - I_{1\perp}(\varphi) = (a_z^4 - a_x^4)\cos 2\varphi + (a_z^2 - a_x^2)^2 \cos^2 2\varphi +$$
$$+ 2(a_x^2 + a_z^2)a_x a_z \sin 2\varphi \cos \delta d + (a_z^2 - a_x^2)a_x a_z \sin 4\varphi \times$$
$$\times (2 + \cos \delta d) + 2a_x^2 a_z^2 \sin^2 2\varphi \cos \delta d, \tag{20}$$

$$I_{1\parallel}(\varphi) + I_{1\perp}(\varphi) = (a_x^2 + a_z^2)^2 + (a_z^4 - a_x^4)\cos 2\varphi. \tag{21}$$

Since in this case the quantity δd occurs only in the expression for the difference $I_{1\parallel}(\varphi) - I_{1\perp}(\varphi)$ and then only in the cosine, the value of $P(\varphi)$ varies strictly periodically with δd. This is due to the fact that in the case under consideration luminescence is excited only in a part of a crystal, and the remainder of the crystal simply acts as a birefringent transparent plate.

§ 3. Polarization Diagrams of a Rotator

It is known from classical electrodynamics that a dipole may emit radiation not only as a result of periodic variation of the absolute value of its dipole moment, but also as a result of rotation. Such a rotating radiation-emitting dipole is known as a rotator. In quantum mechanics the radiation emitted by a linear oscillator and a rotator is treated in the same dipole approximation. The oscillator and rotator simply represent two extreme cases. We shall not consider the general case but we shall see how light is absorbed and emitted by a rotator whose axis is oriented in an arbitrary manner relative to the direction of observation.

We shall assume that a unit vector parallel to the rotator axis has the components a_x, a_y, and a_z (we shall keep the system of coordinates used in discussing a linear oscillator), and we shall postulate that the rotator is illuminated with elliptically polarized light whose electric vector $\mathbf{E}_e$ is described by Eq. (2). The rotator can be represented by two mutually perpendicular linear oscillators whose oscillations are shifted in phase by 90°. We shall select them so that one of them lies in the horizontal plane and the other, in the vertical plane (Fig. 3). Then, the orientation of the former is described by a unit vector $\mathbf{a}_1$ with the coordinates

$$-a_y / \sqrt{a_x^2 + a_y^2}, \quad a_x / \sqrt{a_x^2 + a_y^2}, \quad \text{and } 0,$$

and the latter, by a vector $\mathbf{a}_2$ with the coordinates

$$-a_x a_z / \sqrt{a_x^2 + a_y^2}, \quad -a_y a_z / \sqrt{a_x^2 + a_y^2}, \quad \text{and } \sqrt{a_x^2 + a_y^2}.$$

We shall calculate the rotator energy W_{rot} by finding, as before, the mean-square value of the projection of the electric field on the component linear oscillators. However, since the

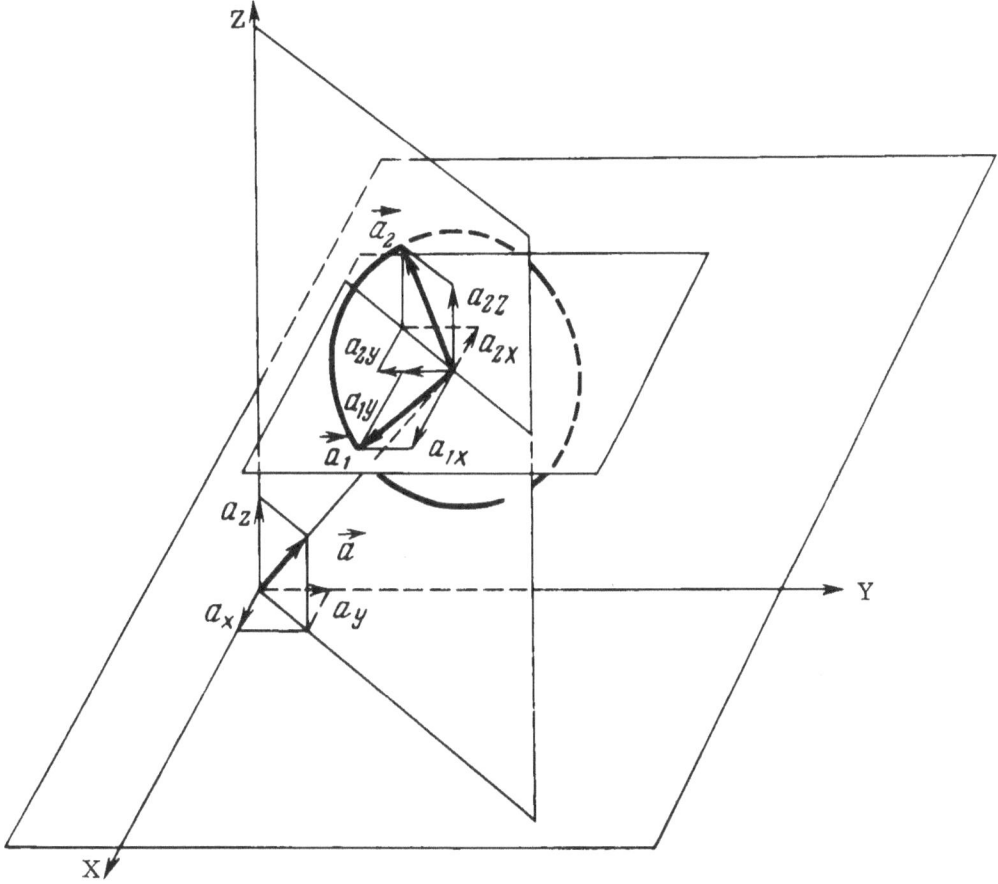

Fig. 3. Representation of a rotator by two coherent oscillators. Here, **a** is a unit vector directed along the rotator axis; a_x, a_y, and a_z are the components of this vector; a_1 and a_2 are unit vectors along the directions of two coherent oscillators into which the rotator is expanded; a_{1x}, a_{1y}, a_{1z}, a_{2x}, a_{2y}, and a_{2z} are the components of these vectors. The circle is the trajectory of the end of the unit vector representing the direction of a rotating dipole.

oscillations are shifted in phase by 90°, this shift must be included in the effective phase of the electric field. We then obtain

$$W_{\text{rot}} \propto \left[\cos \varphi \sqrt{a_x^2 + a_y^2} \cos (\omega t + \delta_e y) - \sin \varphi \frac{a_x a_z}{\sqrt{a_x^2 + a_y^2}} \cos \omega t \pm \sin \varphi \frac{a_y}{\sqrt{a_x^2 + a_y^2}} \sin \omega t \right]^2 \sim$$

$$\sim a_y^2 + a_x^2 \cos^2 \varphi + a_z^2 \sin^2 \varphi - 2a_x a_z \sin \varphi \cos \varphi \cos \delta_e y \mp 2a_y \sin \varphi \cos \varphi \sin \delta_e y. \tag{22}$$

The double sign in front of the last term in the above formula means that the rotator can be right- or left-handed, i.e., it may rotate clockwise and counterclockwise.

In calculating the radiation emitted by a rotator we can also represent it by two coherent linear oscillators, which oscillate with a phase shift of 90°. It is convenient to use the same component oscillators as in the calculation of the absorption. Luminescence is governed by the projections of these oscillators onto a plane which is perpendicular to the direction of observation, i.e., onto the XZ plane. Hence, it follows that the luminescence emitted by a rotator has an electric vector $\mathbf{E}'_l$ with the components

$$\begin{aligned}
E'_{l\mathrm{x}} &\propto \sqrt{W_{\mathrm{rot}}}\left(-\frac{a_x a_z}{\sqrt{a_x^2 + a_y^2}}\cos\omega t \pm \frac{a_y}{\sqrt{a_x^2 + a_y^2}}\sin\omega t\right) \\
E'_{l\mathrm{z}} &\propto \sqrt{W_{\mathrm{rot}}}\sqrt{a_x^2 + a_y^2}\cos\omega t.
\end{aligned}\tag{23}$$

An additional phase shift, equal to $\delta_l\,(d - y)$, appears between these components when they leave the crystal plate:

$$\begin{aligned}
E_{l\mathrm{x}} &\propto \sqrt{W_{\mathrm{rot}}}\left(-\frac{a_x a_z}{\sqrt{a_x^2 + a_y^2}}\cos\omega t \pm \frac{a_y}{\sqrt{a_x^2 + a_y^2}}\sin\omega t\right), \\
E_{l\mathrm{z}} &\propto \sqrt{W_{\mathrm{rot}}}\sqrt{a_x^2 + a_y^2}\cos[\omega t + \delta_\Pi(d - y)].
\end{aligned}\tag{24}$$

The intensity i_1 of the light transmitted by an analyzer, which passes only a flux with an electric vector oriented at an angle ψ with respect to the Z axis, is proportional to

$$\begin{aligned}
i_1 &\propto W_{\mathrm{rot}}\,\overline{(E_{l\mathrm{x}}\sin\psi + E_{l\mathrm{z}}\cos\psi)^2} \propto (a_y^2 + a_x^2\cos^2\varphi + a_z^2\sin^2\varphi - \\
&\quad - 2a_x a_z\sin\varphi\cos\varphi\cos\delta_\mathrm{e}y \mp 2a_y\sin\varphi\cos\varphi\sin\delta_\mathrm{e}y)\,[a_y^2 + a_x^2\cos^2\psi + \\
&\quad + a_z^2\sin^2\psi - 2a_x a_z\sin\psi\cos\psi\cos\delta_l\,(d - y)\mp 2a_y\sin\psi\cos\psi\sin\delta_l(d - y)].
\end{aligned}\tag{25}$$

Here, we have used W_{rot} taken from Eq. (22).

Both factors in i_1 include terms with double signs describing right- and left-handed rotators. If these rotators are present in equal concentrations in a crystal, the summation over all the rotators with a given orientation results in the canceling of all these terms and only the products remain. Therefore, the expression for i_1 can be written in the form

$$\begin{aligned}
i_1 &\propto (a_y^2 + a_x^2\cos^2\varphi + a_z^2\sin^2\varphi - 2a_x a_z\sin\varphi\cos\varphi\cos\delta_\mathrm{e}y)\times \\
&\quad \times [a_y^2 + a_x^2\cos^2\psi + a_z^2\sin^2\psi - 2a_x a_z\sin\psi\cos\psi\cos\delta_l\,(d - y)] + \\
&\quad + 4a_y^2\sin\varphi\cos\varphi\sin\psi\cos\psi\sin\delta_\mathrm{e}y\sin\delta_l(d - y).
\end{aligned}\tag{26}$$

We can now readily calculate $I_{1\parallel}(\varphi)$ and $I_{1\perp}(\varphi)$. As before, we shall integrate i_1 with respect to dy, and we shall substitute into the resultant expression $\psi = \varphi$ for $I_{1\parallel}(\varphi)$ and $\psi = \varphi + \pi/2$ for $I_{1\perp}(\varphi)$. Assuming that $\delta_\mathrm{e} = \delta_l = \delta$, we finally obtain

$$\begin{aligned}
I_{1\parallel}(\varphi) - I_{1\perp}(\varphi) &= \tfrac{1}{2}(1 + a_y^2)(a_x^2 - a_z^2)\cos 2\varphi + \tfrac{1}{2}(a_x^2 - a_z^2)^2\cos^2 2\varphi - \\
&\quad - (1 + a_y^2)a_x a_z\sin 2\varphi\,\frac{\sin\delta d}{\delta d} + (a_z^2 - a_x^2)a_x a_z\sin 4\varphi\,\frac{\sin\delta d}{\delta d} + \\
&\quad + a_x^2 a_z^2\sin^2 2\varphi\left(\frac{\sin\delta d}{\delta d} + \cos\delta d\right) + a_y^2\sin^2 2\varphi\left(\frac{\sin\delta d}{\delta d} - \cos\delta d\right),
\end{aligned}\tag{27}$$

$$I_{1\parallel}(\varphi) + I_{1\perp}(\varphi) = \tfrac{1}{2}(1 + a_y^2) + \tfrac{1}{2}(1 + a_y^2)(a_x^2 - a_z^2)\cos 2\varphi - (1 + a_y^2)a_x a_z\sin 2\varphi\,\frac{\sin\delta d}{\delta d}.\tag{28}$$

These formulas resemble the corresponding formulas (8) and (9). They differ only in respect of some coefficients in front of the trigonometric functions of the angle 2φ and by a free term. Therefore, the general comments made in §1 on the properties of the polarization diagrams, such as their asymmetry, absence of dependence of the values of $P(0)$ and $P(90°)$ on δd, etc., apply also in the case of a rotator. The most important difference between the linear oscillator and rotator cases is as follows. In Eq. (8) the coefficient in front of $\cos 2\varphi$ is smaller (in the absolute sense) than the coefficient in front of $\cos^2 2\varphi$, which is positive. Therefore, we always have $P_{\mathrm{osc}}(0) > 0$ and $P_{\mathrm{osc}}(90°) > 0$. However, in Eq. (27) the coefficient in front of $\cos 2\varphi$ is larger (in the absolute sense) than the coefficient in front of $\cos^2 2\varphi$. This inequality should remain after summation over all equivalent rotators. We then have $P_{rot}(0) < 0$ or $P_{rot}(90°) < 0$. Since at these points the value of P is independent of the birefringence, we can safely conclude that we are dealing with a rotator if it is found experimentally that at one of these points $P < 0$. However, if at both points we have $P > 0$, we cannot draw this conclusion without a more careful quantitative comparison of the theory and experiment.

§ 4. General Case: Combination of a Linear Oscillator with a Rotator

In general, an elementary radiator can be represented in the dipole approximation by a combination of a linear oscillator and a rotator which are almost exactly synchronized with one another. The net dipole moment vector then describes an ellipse in a plane inclined relative to the common axis of the dipole and rotator. If a radiator of this kind has an overall axial symmetry, the position of this ellipse cannot be fixed (because this would reduce the symmetry of the radiator as a whole) but it should precess about this axis. This is equivalent to some departure from synchronization between the linear oscillator and the rotator, i.e., a departure from coherence between them. The departure may be considerable if during the lifetime of an excited state the ellipse makes several revolutions. A difference of only $10^{-4}\%$ between the rotator and oscillator frequencies is sufficient for this to happen, and such a difference fits easily within the width of the lines found for crystals. Therefore, we may assume that in all real cases there is no coherence. It does not follow that there is no exchange of energy between the oscillator and rotator. On the contrary, since this division into an oscillator and a rotator is purely a thought process, we may assume that the excitation energy is distributed between them irrespective of whether it is absorbed by the oscillator or by the rotator.

Clearly, a description of a general dipole radiator must include some characteristic of the ratio of the oscillator and rotator contributions. We shall use the relative contribution of the rotator β and assume that the oscillator contribution is always unity. If $\beta = 0$, we have a linear oscillator, but if $\beta \to \infty$, we have a rotator; if $\beta = 1$, the contributions are equal. The energy of such a general dipole radiator is equal to the sum of the oscillator and rotator energies.

We shall follow the same approach in considering the light emitted by a dipole radiator as representing the emission of light from a linear oscillator and a rotator which are noncoherently coupled. Therefore, we have to add not the electric fields, but the intensities of the emitted light.

If we denote the relative contribution of the rotator to the absorption of the exciting light by β_e and the contribution to the luminescence by β_l, we find from Eqs. (4) and (22) that

$$W_{rad} = W_{osc} + \beta W_{rot} \sim \beta_e + (1-\beta_e)(a_x^2 \sin^2\varphi + a_z^2 \cos^2\varphi + 2a_x a_z \sin\varphi \times \\ \times \cos\varphi \cos\delta_e y) \mp 2\beta_e a_y \sin\varphi \cos\varphi \sin\delta_e y, \tag{29}$$

$$i_1 \propto [\beta_e + (1-\beta_e)(a_x^2 \sin^2\varphi + a_z^2 \cos^2\varphi + 2a_x a_z \sin\varphi \cos\varphi \cos\delta_e y) \mp \\ \mp 2\beta_e a_y \sin\varphi \cos\varphi \sin\delta_e y] \{\beta_l + (1-\beta_l)[a_x^2 \sin^2\psi + a_z^2 \cos^2\psi + \\ + 2a_x a_z \sin\psi \cos\psi \cos\delta_l(d-y)] \mp 2\beta_l a_y \sin\psi \cos\psi \sin\delta_l(d-y)\}. \tag{30}$$

Hence, following the previous procedure we can calculate $I_{1\parallel}(\varphi)$ and $I_{1\perp}(\varphi)$, as well as their sum and difference. We shall give the final result for the case when $\beta_e = \beta_l = \beta$ and $\delta_l = \delta_e = \delta$:

$$I_{1\parallel}(\varphi) - I_{1\perp}(\varphi) = \frac{1}{2}(1-\beta)(a_z^2 - a_x^2)[2\beta + (1-\beta)(a_x^2 + a_z^2) + \\ + (1-\beta)(a_z^2 - a_x^2)\cos 2\varphi]\cos 2\varphi + (1-\beta)a_x a_z[2\beta + \\ + (1-\beta)(a_x^2 + a_z^2) + 2(1-\beta)(a_z^2 - a_x^2)\cos 2\varphi]\sin 2\varphi \frac{\sin\delta d}{\delta d} + \\ + (1-\beta)^2 a_x^2 a_z^2 \sin^2 2\varphi \left(\frac{\sin\delta d}{\delta d} + \cos\delta d\right) + \beta^2 a_y^2 \sin^2 2\varphi \left(\frac{\sin\delta d}{\delta d} - \cos\delta d\right), \tag{31}$$

$$I_{1\parallel}(\varphi) + I_{1\perp}(\varphi) = \frac{1}{2}[2\beta + (1-\beta)(a_x^2 + a_z^2) + (1-\beta)(a_z^2 - a_x^2)\cos 2\varphi] \times$$

$$\times [2\beta + (1-\beta)(a_x^2 + a_z^2)] + (1-\beta)a_x a_z [2\beta + (1-\beta)(a_x^2 + a_z^2)]\sin 2\varphi \frac{\sin \delta d}{\delta d}. \qquad (32)$$

If $\beta = 0$, these formulas reduce, respectively, to Eqs. (8) and (9), and the coefficients in front of β^2 correspond to formulas (27) and (28) [only these terms remain in the expression for $P(\varphi)$ in the limit $\beta \to \infty$]. If $\beta = 1$, i.e., in the case of equal contributions of a linear oscillator and a rotator, Eqs. (31) and (32) simplify considerably to just one term each. Then, the sum $I_{1\parallel} + I_{1\perp}$ is found to be independent of φ. We thus obtain

$$P(\varphi) = \frac{1}{2n}\sin^2 2\varphi \left(\frac{\sin \delta d}{\delta d} - \cos \delta d\right) \sum_{j=1}^{n} a_{\nu j}^2. \qquad (33)$$

It is clear from Eq. (33) that $P(\varphi)$ touches the abscissa only at the points $\varphi = 0$ and $90°$, and it does not intersect it anywhere; moreover, $|P|$ reaches its maximum at $\varphi = 45$ and $135°$. The sign of $P(\varphi)$ is the same as the sign of the difference $(\sin \delta d)/\delta d - \cos \delta d$. If this difference vanishes, we find that $P(\varphi) = 0$ for any value of φ. For other values of δd the quantity $P(\varphi)$ does not vanish. It is interesting to note that the shape of the polarization diagrams is independent of δd, and only the amplitude varies with δd.

These features of the polarization diagrams in the $\beta = 1$ case may be used as an experimental criterion that indeed we have $\beta = 1$, because the same combination of properties is not obtained in any other case. On the other hand, if it is found experimentally that $P(\varphi) = 0$ for all values of φ, we cannot say that the investigated centers do not have a definite orientation: We must first determine the polarization diagram after altering the thickness of a crystal somewhat in order to ensure that $(\sin \delta d/\delta d) - \cos \delta d \neq 0$. Moreover, we must establish that there is no depolarization of light (this will be discussed in the next section).

It is also clear from Eq. (33) that if $\beta = 1$, the polarization diagram is symmetric relative to $\varphi = 90°$ irrespective of how the radiators are oriented. This demonstrates once again that the symmetry of the polarization diagram is not a proof of the symmetry of radiators relative to the XY plane although in the absence of dichroism an asymmetric polarization diagram definitely indicates an asymmetry in the orientation of the radiators.

§ 5. Influence of Depolarization

In general, there may be three reasons why the experimentally observed polarization ratio is smaller than expected: There may be no fixed preferential direction in each luminescence center; light may be depolarized because of imperfections in a crystal or on its surface; light may be depolarized because of the migration of energy between luminescence centers. These factors have different effects on the polarization diagrams.

The absence of a fixed preferential direction of radiators makes the probability of the absorption of exciting light and the probability of emission of luminescence photons independent of the direction of the electric vector of the incident (exciting) light. Therefore, a set of such radiators emits unpolarized light. This means that $I_\parallel = I_\perp$ and $P(\varphi) = 0$. This case corresponds to a "point" defect (in contrast to the two-point defects discussed in the Introduction) in a crystal field of sufficiently high symmetry. For example, a manganese ion in cubic zinc sulfide may not give rise to a significant luminescence polarization, since it replaces a zinc ion without any compensation, and the radiative transitions occur in the inner electron shell.

The second cause of the reduction in $P(\varphi)$ is in no way related to luminescence centers but it can be allowed for approximately by a suitable correction to the theoretical curves. If, because of imperfection of a crystal (for example, due to scattering of light by defects), the transmitted light becomes partly depolarized, this is equivalent to the addition of a constant correction to $I_\parallel(\varphi)$ and $I_\perp(\varphi)$. In the numerator of the expression for P, these corrections cancel out, but in the denominator they add, and this increases the constant term in the denominator. The correction can be found by measuring, exactly as in the determination of δd, the polarization diagram of light transmitted across a crystal without absorption. As pointed out earlier, in this case we should have P(0) = P(90°) = 1. If the transmitted light is partly depolarized, then at these points we have P < 1 although, as before, they correspond to a maximum of $P(\varphi)$. The value of $1/P_{max}$ shows how much we have to increase the constant term in the formula for the polarization diagram in order to allow for the depolarization of light resulting from the imperfection of a crystal. After this correction the theoretical curve can be compared with the experimental results not only in respect of its general shape, but also in respect of the absolute values of the coordinates.

It is much more difficult to allow quantitatively for the migration of the excitation energy although in some cases it can be done. Therefore, it is best to avoid such migration by selecting suitably the experimental conditions. In the recombination luminescence mechanism this can be done by lowering the temperature and increasing the excitation intensity. If the energy is transferred not as a result of recombination, but by resonance, we have to reduce the activator concentration. Luminescence must be excited in the absorption band of the investigated luminescence centers and not in the absorption region of the host substance or other centers because otherwise the investigated luminescence centers would receive energy only as a result of migration.

We shall now consider how migrational depolarization affects the polarization diagrams. We can do this by calculating W again. The physical meaning of this quantity is that it is proportional to the probability of absorption of an exciting photon by a radiator, i.e., it is proportional to the probability of a transition of the radiator to an excited state. In the presence of energy migration this probability should be supplemented by the probability of a transition of this radiator to an excited state as a result of acquisition of energy by migration from other centers.*

Several migration events of these kinds may cause the radiator which has absorbed a quantum of exciting light to "forget" completely its orientation. In the very widely encountered case of recombination migration, when the energy is transferred as a result of a transition of an electron and a hole to another center, a radiator which has absorbed a quantum of exciting light "forgets" its orientation after the first migration event. If we assume that the migration correction is completely independent of φ, we find that

$$W_{tot} = W_{migr} + W_{osc} + \beta_e W_{rot} \propto \alpha + \beta_e + (a_x^2 \sin^2\varphi + a_z^2 \cos^2\varphi + \\ + 2a_x a_z \sin\varphi \cos\varphi \cos\delta_e y)(1 - \beta_e) \mp 2\beta_e a_y \sin\varphi \cos\varphi \sin\delta_e y. \qquad (34)$$

Here, W_{migr} and α are constants which are proportional to one another but independent of φ and the radiator orientation. Using this formula we can calculate the value of i_1, and then, proceeding as before, we can find the value of $P(\varphi)$. We should bear in mind that the second factor in the expression for i_1 does not contain α because it applies to the luminescence, whose

* Clearly, this increase in the probability of a transition to an excited state does not increase the concentration of excited radiators because of a simultaneous increase in the probability of the loss of excitation energy by the radiator as a result of its transfer to a different radiator. However, this is of little importance in our case because it simply gives rise to a constant factor (smaller than unity) in all terms in the expression for W, and this factor disappears from the final formula for $P(\varphi)$.

emission implies the completion of the energy migration process. We thus obtain (again assuming that $\beta_e = \beta_l = \beta$ and $\delta_l = \delta_e = \delta$)

$$I_{1\parallel}(\varphi) - I_{1\perp}(\varphi) = f_1(\beta, \delta d, a_x, a_y, a_z, \varphi) +$$

$$+ \alpha(1-\beta)\left[(a_z^2 - a_x^2)\cos 2\varphi + 2a_x a_z \sin 2\varphi \frac{\sin \delta d}{\delta d}\right], \qquad (35)$$

$$I_{1\parallel}(\varphi) + I_{1\perp}(\varphi) = f_2(\beta, \delta d, a_x, a_y, a_z, \varphi) + \alpha[2\beta + (1-\beta)(a_x^2 + a_z^2)], \qquad (36)$$

where the functions f_1 and f_2 are the right-hand sides of Eqs. (31) and (32), describing the polarization diagram of a general dipole radiator in the absence of energy migration. Thus, if we substitute $\alpha = 0$, we obtain once again Eqs. (31) and (32).

If in the final expression for $P(\varphi)$ we allow α to go to infinity, we obtain an expression for the polarization diagram in the case of uniform excitation of all luminescence centers of one kind irrespective of their orientation:

$$P_{\text{migr}}(\varphi) = (1-\beta)\frac{\cos 2\varphi \sum_{j=1}^{n}(a_{zj}^2 - a_{xj}^2) + 2\sin 2\varphi \frac{\sin \delta d}{\delta d}\sum_{j=1}^{n} a_{xj}a_{zj}}{2\beta n + (1-\beta)\sum_{j=1}^{n}(a_{xj}^2 + a_{zj}^2)}. \qquad (37)$$

It is clear from Eqs. (36) and (37) that the sum $I_\parallel + I_\perp$ is independent of φ if the migration of excitation energy is important. In this sense the energy migration process has the same result as the equality of the contributions of an oscillator and a rotator [$\beta = 1$, see Eq. (33)]. However, if $\beta = 1$ (and $\alpha = 0$), the polarization diagram differs from that obtained in the case of strong energy migration.

It is clear from Eqs. (37) and (33) that when energy migration is strong, the polarization diagram is a sinusoid with an argument 2φ and a phase which depends on δd. In the absence of migration and if $\beta = 1$, the polarization diagram is the square of this sinusoid, but its phase is independent of δd, and, as mentioned earlier, $P(\varphi) = 0$ only for certain values of δd. However, if $\beta = 1$ in the case of strong energy migration, then $P(\varphi) = 0$ for any δd.

Thus, we cannot say that radiators do not have a fixed preferential orientation simply because $P(\varphi) = 0$ for any thickness of a crystal. We must also show that $\beta \neq 1$ or that excitation energy migration between luminescence centers does not play any significant role.

The importance or otherwise of migrational depolarization can best be established experimentally by measuring the polarization diagram not only at the moment of excitation, but also during the afterglow stage. Since a finite time is required for the transfer of energy from one center to another, it follows that after the end of excitation the role of the luminescence centers which have received energy from other centers gradually increases in importance. Therefore, the parameter α should increase monotonically during afterglow. In the case of resonance energy migration, this increase should be exponential. We shall now prove this prediction.

We shall use n' to denote the concentration of excited centers which have received energy directly from the exciting light and n" to denote the concentration of excited centers which have received energy by migration from other centers. Time dependences of these concentrations after the end of excitation are described by a system of differential equations,

$$\frac{dn'}{dt} = \mathcal{I} - n'(r' + r_0), \qquad \frac{dn''}{dt} = n'r' - n''r_0, \qquad (38)$$

where $\mathcal{I}$ is the number of times that the quanta of the exciting light are absorbed per unit volume and per unit time; r' is the probability of the transfer of energy from one center to another; r_0 is the sum of the probabilities of emission of a light quantum and of the conversion of the excitation energy of a given center into heat. The solution of the system is of the form

$$
\left.
\begin{aligned}
n' &= n'_0 \exp\left[-(r'+r_0)t\right], \\
n'' &= n'_0 \left\{ \frac{r'+r_0}{r_0}\exp(-r_0 t) - \exp\left[-(r'+r_0)t\right] \right\},
\end{aligned}
\right\} \tag{39}
$$

where n'_0 is the value of n' at the end of excitation (we shall assume that the dynamic equilibrium has been established between the n' and n" centers before the end of excitation).

The quantity α in Eqs. (35) and (36) is, by definition, the ratio of n" to n'. Therefore,

$$
\alpha \equiv \frac{n''}{n'} = \left(1 + \frac{r'}{r_0}\right)e^{r't} - 1, \tag{40}
$$

which was the point that we set out to prove.

In the case of energy migration due to recombination this simple dependence $\alpha(t)$ does not apply but a monotonic rise of α still occurs.

It is not easy to find α from the experimental results: One has to select from a two-parameter family of the polarization diagrams (parameters α and β) that diagram which agrees best with the experimental curve. Since on top of that we do not know the radiator orientation such a selection is likely to be ambiguous. This ambiguity can be reduced by using the following procedure.

It is clear from Eqs. (35), (36), (31), and (32) that the ratios

$$
S_1 \equiv \frac{[I_\parallel(0)-I_\perp(0)]+[I_\parallel(90°)-I_\perp(90°)]}{[I_\parallel(0)+I_\perp(0)]-[I_\parallel(90°)+I_\perp(90°)]} = \frac{\displaystyle\sum_{j=1}^{n}(a_{zj}^2-a_{xj}^2)^2}{\displaystyle\sum_{j=1}^{n}(a_{zj}^4-a_{xj}^4)}, \tag{41}
$$

$$
S_2 \equiv \frac{I_\parallel(0)-I_\perp(90°)}{I_\parallel(90°)-I_\perp(0)} = -\frac{\displaystyle\sum_{j=1}^{n}a_{zj}^2(a_{zj}^2-a_{xj}^2)}{\displaystyle\sum_{j=1}^{n}a_{xj}^2(a_{zj}^2-a_{xj}^2)} \tag{42}
$$

are independent of α, β, and δd. Therefore, we can calculate these quantities from the experimental results and compare them with the theoretical values for various possible radiator orientations, and we can thus select the most suitable ratio. When this is done, the problem is solved. There should be no difficulty in the subsequent calculation of α and β. This can be done by defining a third ratio S_3:

$$
S_3 \equiv \frac{I_\perp(0)-I_\perp(90°)}{I_\parallel(90°)-I_\perp(0)} = \frac{\alpha+\beta}{1-\beta}\frac{\displaystyle\sum_{j=1}^{n}(a_{xj}^2-a_{zj}^2)}{\displaystyle\sum_{j=1}^{n}a_{xj}^2(a_{xj}^2-a_{zj}^2)}. \tag{43}
$$

If a_{xj} and a_{zj} are known, we can find

$$
q \equiv \frac{\alpha+\beta}{1-\beta} = S_3 \frac{\displaystyle\sum_{j=1}^{n}a_{xj}^2(a_{xj}^2-a_{zj}^2)}{\displaystyle\sum_{j=1}^{n}(a_{xj}^2-a_{zj}^2)}. \tag{44}
$$

Substituting q in the expressions for P(0) and P(90°) obtained from Eqs. (35) and (36), we obtain

$$P(0) = \frac{q \sum\limits_{j=1}^{n} (a_{zj}^2 - a_{xj}^2) + \sum\limits_{j=1}^{n} a_{zj}^2 (a_{zj}^2 - a_{xj}^2)}{q \dfrac{2\beta}{1-\beta} + q \sum\limits_{j=1}^{n} (a_{xj}^2 + a_{zj}^2) + \sum\limits_{j=1}^{n} a_{zj}^2 (a_{xj}^2 + a_{zj}^2) + \dfrac{2\beta}{1-\beta} \sum\limits_{j=1}^{n} a_{zj}^2}, \tag{45}$$

$$P(90°) = \frac{-q \sum\limits_{j=1}^{n} (a_{zj}^2 - a_{xj}^2) + \sum\limits_{j=1}^{n} a_{xj}^2 (a_{zj}^2 - a_{xj}^2)}{q \dfrac{2\beta}{1-\beta} + q \sum\limits_{j=1}^{n} (a_{xj}^2 + a_{zj}^2) + \sum\limits_{j=1}^{n} a_{xj}^2 (a_{xj}^2 + a_{zj}^2) + \dfrac{2\beta}{1-\beta} \sum\limits_{j=1}^{n} a_{xj}^2}. \tag{46}$$

Each of the above equations can be used to find β if we know q and all values of a_{xj} and a_{zj}. This allows us to check once again the correctness of the selected radiator orientation. When we know β and q, we can readily calculate α with the aid of Eq. (44).

The main source of possible errors is in the selection of the radiator orientation. Therefore, it is useful to recheck the correctness of the selected orientation. This can be done by rotating a crystal through 90° about its optic axis (or by cutting a new crystal plate oriented in a suitable manner) and then repeating measurements of the polarization diagram. Such rotation results in a transposition of the squares of the radiator coordinates a_x^2 and a_z^2 in Eqs. (41)-(45). Therefore, the quantities S_1, S_2, S_3, P(0), and P(90°) can be calculated for the new position of the crystal in advance and compared with the values found experimentally. Finally, measurements of the polarization diagram can be repeated during various stages of the afterglow. As mentioned earlier, only the value of α should increase with time and this can easily be checked experimentally.

Thus, in spite of migrational depolarization, we can find the radiator orientation and the ratio of the contributions of an oscillator and a rotator, as well as the rate of energy migration.

It should be noted that this method for finding the radiator orientation can be used also in the absence of energy migration. It is convenient because we do not have to know the value of δd and Kd (if there is no dichroism).

§6. Analysis of Some Specific Cases

The above theory can be applied, in particular, in studies of an extensive class of crystals whose lattice resembles that of "cubic" zinc sulfide. Such crystals exhibit relatively weak birefringence. This allows us to study fairly thick plates, which is quite convenient from several points of view, especially from the point of view of the total luminescence flux. The structure of these crystals can be regarded as cubic with a certain number of stacking faults. From a crystallographic point of view a part of the lattice following a stacking fault is apparently rotated by 180° relative to the part preceding this fault. Therefore, the whole crystal can be regarded as consisting entirely of cubic lattice layers rotated relative to one another through 180°. The boundaries between these layers are formed by stacking faults. If we ignore changes in the interatomic spacing in the stacking fault regions, the orientation in such crystals can be regarded as identical with the orientation of radiators in crystals with a purely cubic lattice. However, each stacking fault produces a distribution of the adjacent interatomic layers corresponding to a hexagonal structure. Hexagonal interlayers of this kind may occupy a small fraction of a crystal but they give rise to considerable birefringence.

In calculations of the polarization diagrams of such a crystal an allowance for stacking faults has to be made for two reasons: 1) because of the birefringence due to the presence of these faults; 2) because of the resultant rotation of the crystal lattice. However, this rotation does not always affect the radiator orientation and we shall at this stage ignore its influence.

We calculated the polarization diagrams for the following orientations of the radiator axes:

1. Parallel to Twofold Symmetry Axes of the Cubic Lattice of Zinc Sulfide. This direction corresponds to a line joining two nearest like lattice sites. Radiators can be oriented in this way if they represent pairs of impurity atoms (ions) or lattice defects occupying the same sites. Examples of such pairs are an oxygen ion and a sulfur vacancy, two oxygen ions, copper and aluminum ions.

2. Parallel to Threefold Symmetry Axes of the Cubic Lattice. This direction corresponds to a line joining two nearest unlike lattice sites. Examples are centers consisting of an oxygen ion and a zinc vacancy, copper and chlorine ions, etc.

3. Parallel to Fourfold Symmetry Axes of the Cubic Lattice. This direction corresponds to a line joining a lattice site to the nearest tetrahedral void. Examples are centers consisting of a zinc vacancy and a zinc ion in the next interstice or two copper ions, one at a site and the other in an interstice, etc.

For all three orientations we considered the cases when the direction of observation was parallel to the edge of the tetrahedron formed by the sulfur atoms (in zinc sulfide crystals) and when this direction was in the plane of symmetry of this tetrahedron. In all six cases the formulas for the polarization diagrams were derived for the following variants: 1) a linear oscillator absorbs and emits luminescence ($\beta_e = \beta_l = 0$); 2) a linear oscillator absorbs and a rotator emits ($\beta_e = 0$, $\beta_l = \infty$); 3) a rotator absorbs and a linear oscillator emits ($\beta_e = \infty$, $\beta_l = 0$) (the last two cases are indistinguishable); 4) a rotator absorbs and emits ($\beta_e = \beta_l = \infty$); 5) a dipole radiator absorbs and emits ($\beta_e = \beta_l$). Tables 1-3 give the formulas for all these cases. The corresponding families of the polarization curves are plotted in Figs. 4-9. We can see that there is a great variety of these diagrams (particularly if the sign of P is allowed for) and that many cases are easily distinguishable. For example, when radiators are oriented along the fourfold axes we always have P(0) = 0, but when they are oriented along the twofold or threefold axes, there is an extremum at this point or close to it. The case when linear oscillators absorb and emit can be easily distinguished from all the other cases. Thus, if linear oscillators are oriented along the twofold or threefold axes, the values of P are positive for the majority of the angles φ and values of δd (Fig. 4), and, moreover, at the points $\varphi = 0$ and 90° these values reach 60%, which is considerably greater than the value of P for all the other models of the centers. However, if linear oscillators are oriented along the fourfold axes, we find that P(90°) > 0 [P(0°) = 0] and, amounting to about 20%, they are over four times as large as P(90°) for any other orientation of centers along the fourfold axes.

If linear oscillators absorb and rotators emit or, conversely, if rotators absorb and linear oscillators emit, then the orientation of the centers along the twofold or threefold axes is characterized by largely negative values of $P(\varphi)$, as shown in Fig. 5, and this applies also to $\varphi = 0$ and 90°, whereas near 0 and 90° there are minima and not maxima. In the case of orientation along the fourfold axes the values of P(90°) are again negative [P(0°) = 0].

The most difficult to distinguish are the orientations of radiators along the twofold and threefold axes because the corresponding polarization diagrams differ only quantitatively. The task can be simplified by using the values of S_1 and S_2 for all three orientations discussed above (Table 4). We can see that they differ quite considerably.

It is very easy to distinguish a model with a symmetric distribution of oscillators relative to the direction of observation from an asymmetric distribution (it is assumed that the direc-

TABLE 1. Polarization Diagram Formulas for Absorption and Emission by Linear Oscillators

Radiator orientation	Direction of observation relative to tetrahedron symmetry plane	$P(\varphi)$
Parallel to twofold symmetry axes	Parallel	$\dfrac{\cos 2\varphi + 23\cos^2 2\varphi + 8\sin^2 2\varphi\left(\dfrac{\sin \delta d}{\delta d} + \cos \delta d\right)}{39 + \cos 2\varphi}$
	Perpendicular	$\dfrac{\cos 2\varphi + 23\cos^2 2\varphi - 2\sqrt{2}\sin 4\varphi\,\dfrac{\sin \delta d}{\delta d} + 8\sin^2 2\varphi\left(\dfrac{\sin \delta d}{\delta d} + \cos \delta d\right)}{39 + \cos 2\varphi - 2\sqrt{2}\sin 2\varphi\,\dfrac{\sin \delta d}{\delta d}}$
Parallel to threefold axes	Parallel	$\dfrac{\cos 2\varphi + 11\cos^2 2\varphi + 2\sin^2 2\varphi\left(\dfrac{\sin \delta d}{\delta d} + \cos \delta d\right)}{15 + \cos 2\varphi}$
	Perpendicular	$\dfrac{\cos 2\varphi + 11\cos^2 2\varphi - 2\sqrt{2}\sin 4\varphi\,\dfrac{\sin \delta d}{\delta d} + 2\sin^2 2\varphi\left(\dfrac{\sin \delta d}{\delta d} + \cos \delta d\right)}{15 + \cos 2\varphi + 2\sqrt{2}\sin 2\varphi\,\dfrac{\sin \delta d}{\delta d}}$
Parallel to fourfold axes	Parallel	$\dfrac{-\cos 2\varphi + \cos^2 2\varphi + 4\sin^2 2\varphi\left(\dfrac{\sin \delta d}{\delta d} + \cos \delta d\right)}{9 - \cos 2\varphi}$
	Perpendicular	$\dfrac{-\cos 2\varphi + \cos^2 2\varphi - 2\sqrt{2}\sin 4\varphi\,\dfrac{\sin \delta d}{\delta d} + 4\sin^2 2\varphi\left(\dfrac{\sin \delta d}{\delta d} + \cos \delta d\right)}{9 - \cos 2\varphi - 2\sqrt{2}\sin 2\varphi\,\dfrac{\sin \delta d}{\delta d}}$

TABLE 2. Polarization Diagram Formulas for Absorption by Linear Oscillators and Emission from Rotators, or Absorption by Rotators and Emission from Linear Oscillators

Radiator orientation	Direction of observation relative to tetrahedron symmetry plane	$P(\varphi)$
Parallel to twofold symmetry axes	Parallel	$$\dfrac{-\cos 2\varphi - 23\cos^2 2\varphi - 8\sin^2 2\varphi \left(\dfrac{\sin \delta d}{\delta d} + \cos \delta d\right)}{57 - \cos 2\varphi}$$
	Perpendicular	$$\dfrac{-\cos 2\varphi - 23\cos^2 2\varphi + 2\sqrt{2}\sin 2\varphi \dfrac{\sin \delta d}{\delta d} - 2\sqrt{2}\sin 4\varphi \dfrac{\sin \delta d}{\delta d} - 8\sin^2 2\varphi \left(\dfrac{\sin \delta d}{\delta d} + \cos \delta d\right)}{57 - \cos 2\varphi + 2\sqrt{2}\sin 2\varphi \dfrac{\sin \delta d}{\delta d}}$$
Parallel to threefold axes	Parallel	$$\dfrac{-\cos 2\varphi - 11\cos^2 2\varphi - 2\sin^2 2\varphi \left(\dfrac{\sin \delta d}{\delta d} + \cos \delta d\right)}{21 - \cos 2\varphi}$$
	Perpendicular	$$\dfrac{-\cos 2\varphi - 11\cos^2 2\varphi - 2\sqrt{2}\sin 2\varphi \dfrac{\sin \delta d}{\delta d} + 2\sqrt{2}\sin 4\varphi \dfrac{\sin \delta d}{\delta d} - 2\sin^2 2\varphi \left(\dfrac{\sin \delta d}{\delta d} + \cos \delta d\right)}{21 - \cos 2\varphi - 2\sqrt{2}\sin 2\varphi \dfrac{\sin \delta d}{\delta d}}$$
Parallel to fourfold axes	Parallel	$$\dfrac{\cos 2\varphi - \cos^2 2\varphi - 4\sin^2 2\varphi \left(\dfrac{\sin \delta d}{\delta d} + \cos \delta d\right)}{15 + \cos 2\varphi}$$
	Perpendicular	$$\dfrac{\cos 2\varphi - \cos^2 2\varphi - 2\sqrt{2}\sin 2\varphi \dfrac{\sin \delta d}{\delta d} + 2\sqrt{2}\sin 4\varphi \dfrac{\sin \delta d}{\delta d} - 4\sin^2 2\varphi \left(\dfrac{\sin \delta d}{\delta d} + \cos \delta d\right)}{15 + \cos 2\varphi - 2\sqrt{2}\sin 2\varphi \dfrac{\sin \delta d}{\delta d}}$$

TABLE 3. Polarization Diagram Formulas for Absorption and Emission by Rotators

Radiator orientation	Direction of observation relative to tetrahedron symmetry plane	$P(\varphi)$
Parallel to twofold symmetry axes	Parallel	$$\dfrac{\cos 2\varphi + 23 \cos^2 2\varphi + 56 \sin^2 2\varphi \,\frac{\sin \delta d}{\delta d} - 40 \sin^2 2\varphi \cos \delta d}{135 + \cos 2\varphi}$$
	Perpendicular	$$\dfrac{\cos 2\varphi + 23 \cos^2 2\varphi - 2\sqrt{2}\sin 4\varphi \,\frac{\sin \delta d}{\delta d} + 2\sqrt{2}\sin 4\varphi \,\frac{\sin \delta d}{\delta d} + 56 \sin^2 2\varphi \,\frac{\sin \delta d}{\delta d} - 40 \sin^2 2\varphi \cos \delta d}{135 + \cos 2\varphi - 2\sqrt{2}\sin 2\varphi \,\frac{\sin \delta d}{\delta d}}$$
Parallel to threefold axes	Parallel	$$\dfrac{\cos 2\varphi + 11 \cos^2 2\varphi + 38 \sin^2 2\varphi \,\frac{\sin \delta d}{\delta d} - 34 \sin^2 2\varphi \cos \delta d}{51 + \cos 2\varphi}$$
	Perpendicular	$$\dfrac{\cos 2\varphi + 11 \cos^2 2\varphi + 2\sqrt{2}\sin 2\varphi \,\frac{\sin \delta d}{\delta d} - 2\sqrt{2}\sin 4\varphi \,\frac{\sin \delta d}{\delta d} + 20 \sin^2 2\varphi \,\frac{\sin \delta d}{\delta d} - 16 \sin^2 2\varphi \cos \delta d}{51 + \cos 2\varphi + 2\sqrt{2}\sin 2\varphi \,\frac{\sin \delta d}{\delta d}}$$
Parallel to fourfold axes	Parallel	$$\dfrac{-\cos 2\varphi + \cos^2 2\varphi + 16 \sin^2 2\varphi \,\frac{\sin \delta d}{\delta d} - 8 \sin^2 2\varphi \cos \delta d}{33 - \cos 2\varphi}$$
	Perpendicular	$$\dfrac{-\cos 2\varphi + \cos^2 2\varphi + 2\sqrt{2}\sin 2\varphi \,\frac{\sin \delta d}{\delta d} - 2\sqrt{2}\sin 4\varphi \,\frac{\sin \delta d}{\delta d} + 16 \sin^2 2\varphi \,\frac{\sin \delta d}{\delta d} - 8 \sin^2 2\varphi \cos \delta d}{33 - \cos 2\varphi + 2\sqrt{2}\sin 2\varphi \,\frac{\sin \delta d}{\delta d}}$$

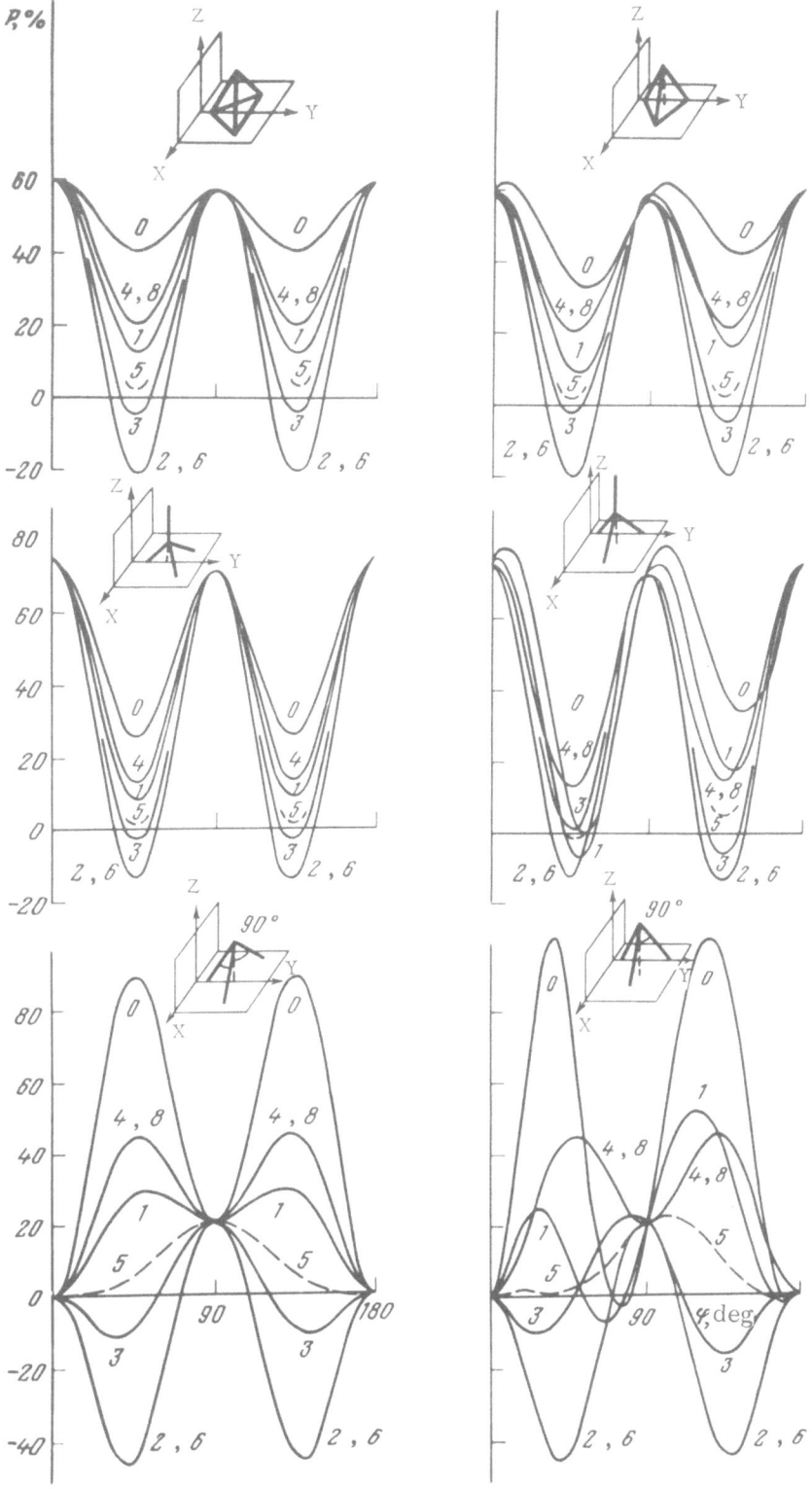

Fig. 4. Polarization diagrams for the case of absorption and emission by a linear oscillator. The diagrams on the left are plotted for the direction of observation Y lying in the symmetry plane of the tetrahedron; the diagrams on the right are plotted for the direction of observation perpendicular to this plane. The upper row of diagrams applies to the orientation of radiators along twofold symmetry axes, the middle row applies to the threefold axes, and the lower one to the fourfold axes; the orientations of the radiators are shown above the polarization diagrams. The numbers alongside the curves give the values of δd in units of $\pi/2$.

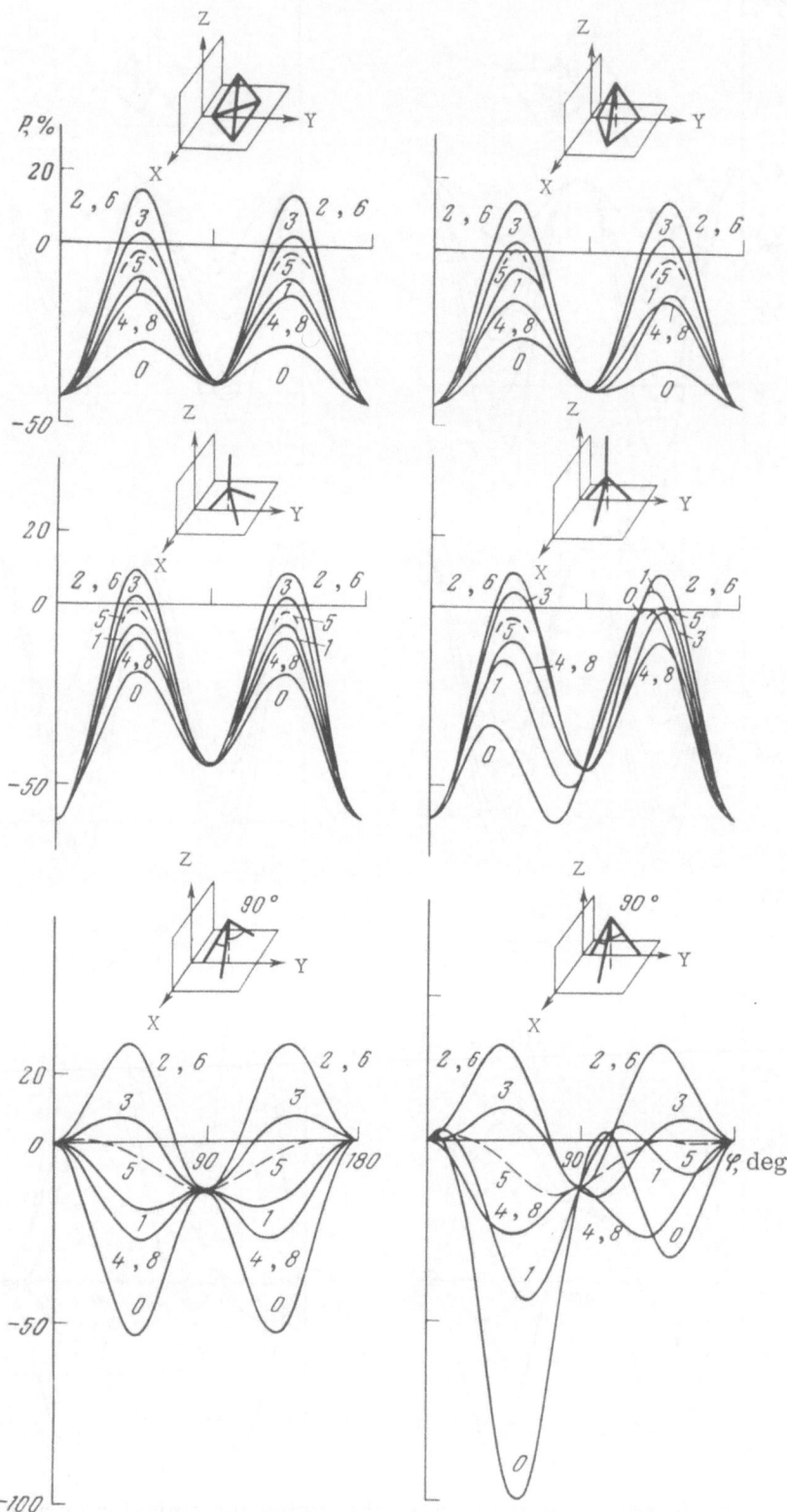

Fig. 5. Polarization diagrams for the case of absorption by a linear oscillator and emission by a rotator, or absorption by a rotator and emission by a linear oscillator. The arrangement and designation of the curves are the same as in Fig. 4.

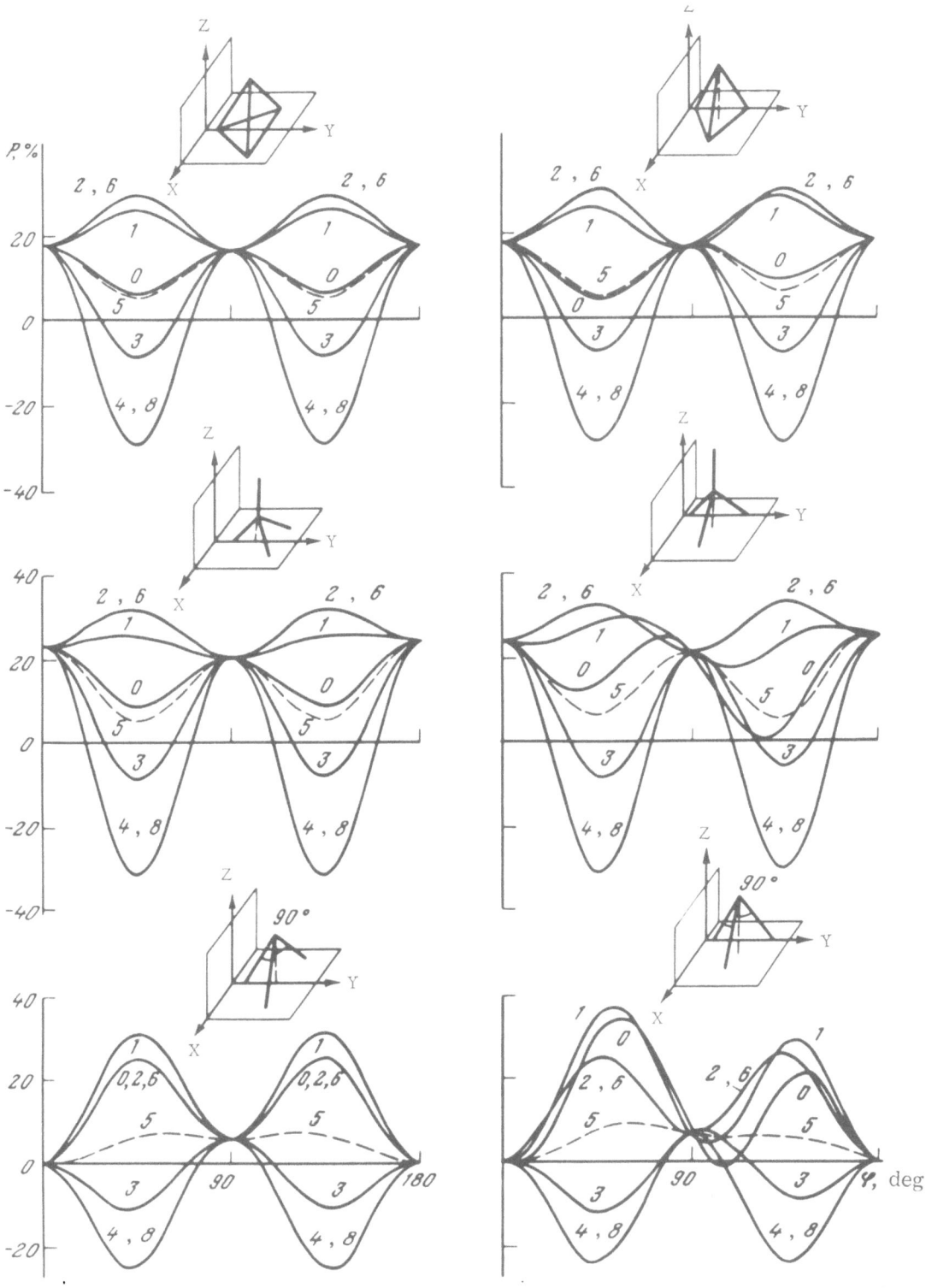

Fig. 6. Polarization diagrams for the case when a rotator both absorbs and emits. The arrangement and designations are the same as in Fig. 4.

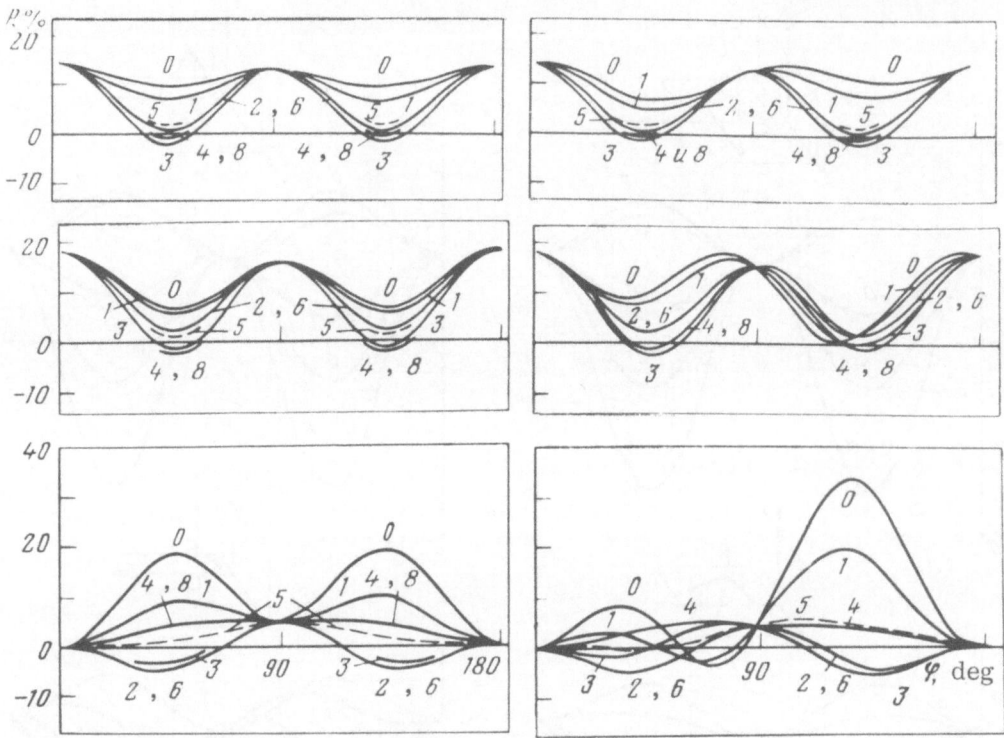

Fig. 7. Polarization diagrams for the case when a general dipole radiator absorbs and emits, and the relative contribution of a rotator is $\beta = 0.3$. The arrangement and designations are the same as in Fig. 4.

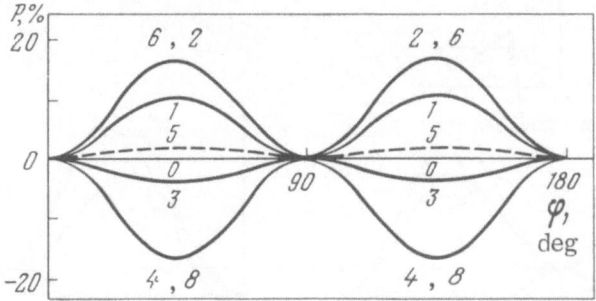

Fig. 8. Polarization diagrams for the case when a general dipole radiator absorbs and emits, and the relative contribution of a rotator is $\beta = 1$. In this case all six radiator orientations have the same polarization diagram.

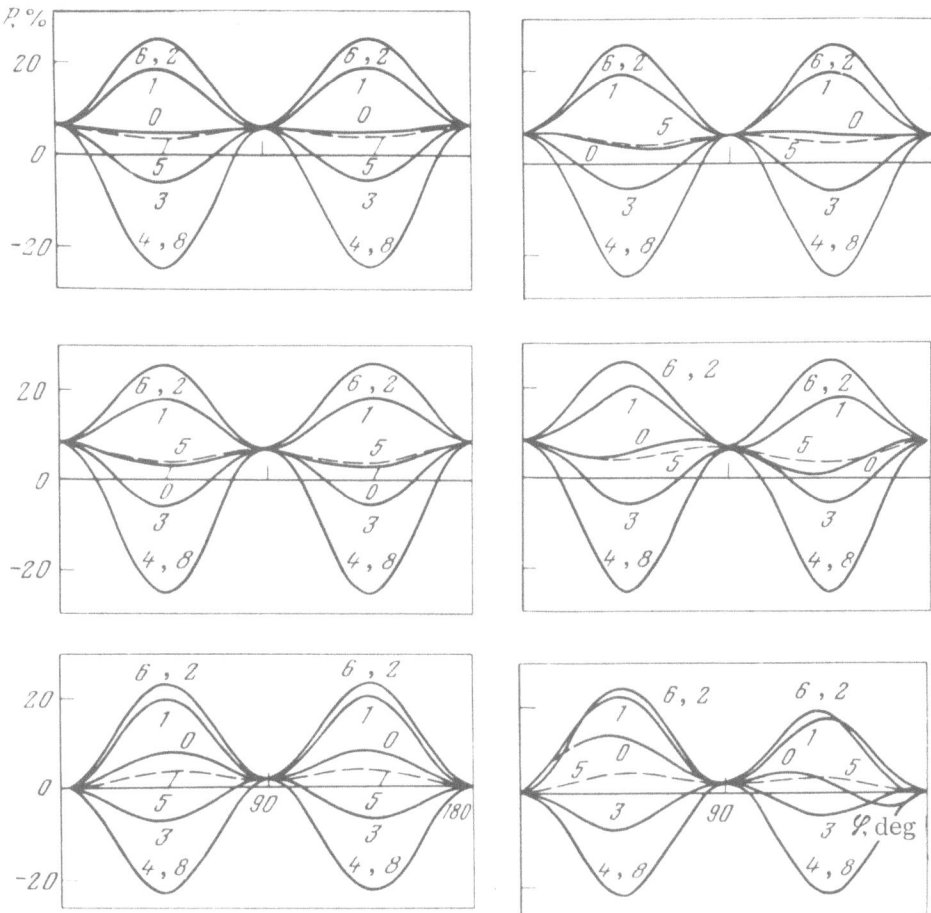

Fig. 9. Polarization diagrams for the case when a general dipole radiator absorbs and emits, and the relative contribution of a rotator is $\beta = 3$. The arrangement and designations are the same as in Fig. 4.

tion of observation is parallel to the edge of the main tetrahedron). An asymmetric polarization diagram should be observed only for an asymmetric position of radiators, and it is manifested most clearly in the model of a general dipole radiator (see Figs. 7-9).

TABLE 4. Values of S_1 and S_2 for Absorption and Emission by Rotators

Radiator orientation	S_1	S_2
Parallel to symmetry axes:		
twofold	+23	$+\dfrac{12}{11}$
threefold	+11	$+\dfrac{6}{5}$
fourfold	−1	0

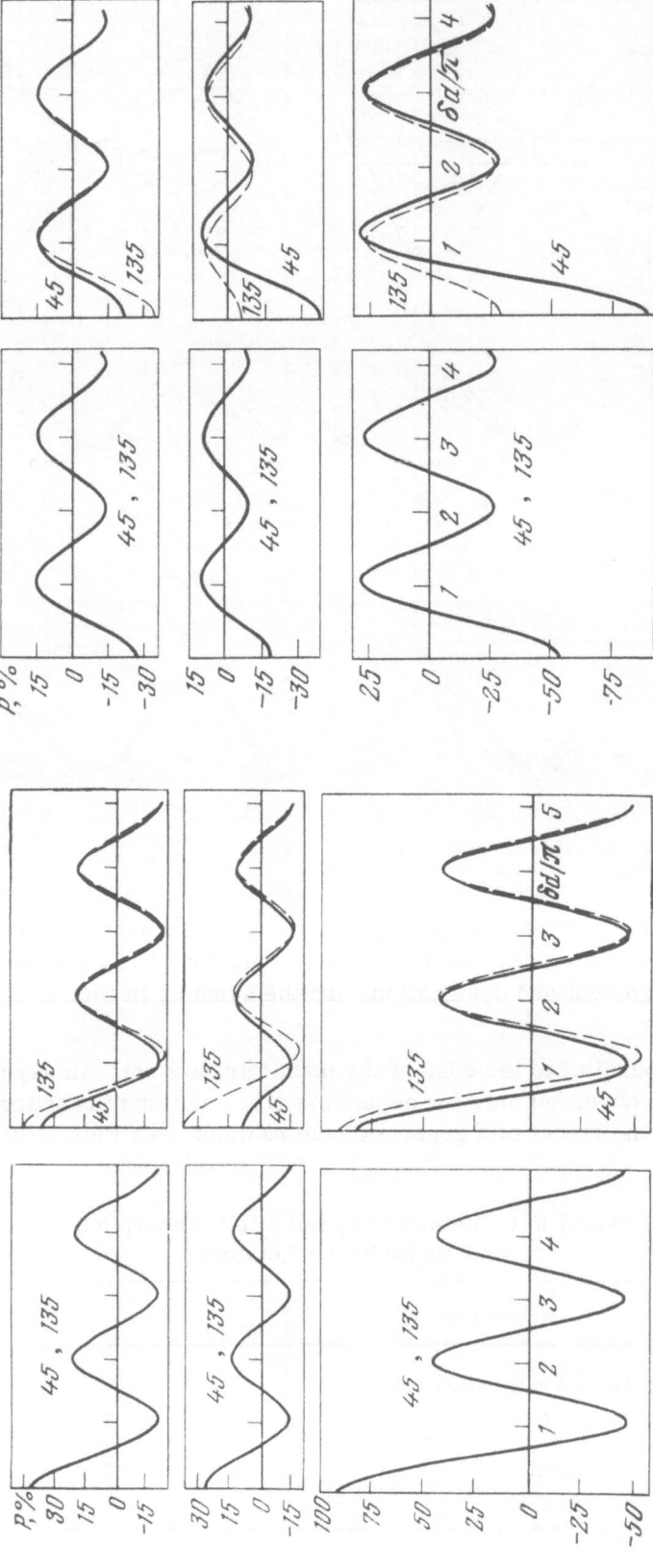

Fig. 10. Dependences of the polarization ratio at the points $\varphi = 45°$ and $135°$ on the value of δd for the case when a linear oscillator absorbs and emits. The numbers alongside the curves are the values of φ in degrees. The arrangement of the curves is the same as in Fig. 4.

Fig. 11. Dependences of the polarization ratio at the points $\varphi = 45°$ and $135°$ on the value of δd for the case when a linear oscillator absorbs and a rotator emits, or when a rotator absorbs and a linear oscillator emits. The numbers alongside the curves are the values of φ in degrees. The arrangement is the same as in Fig. 4.

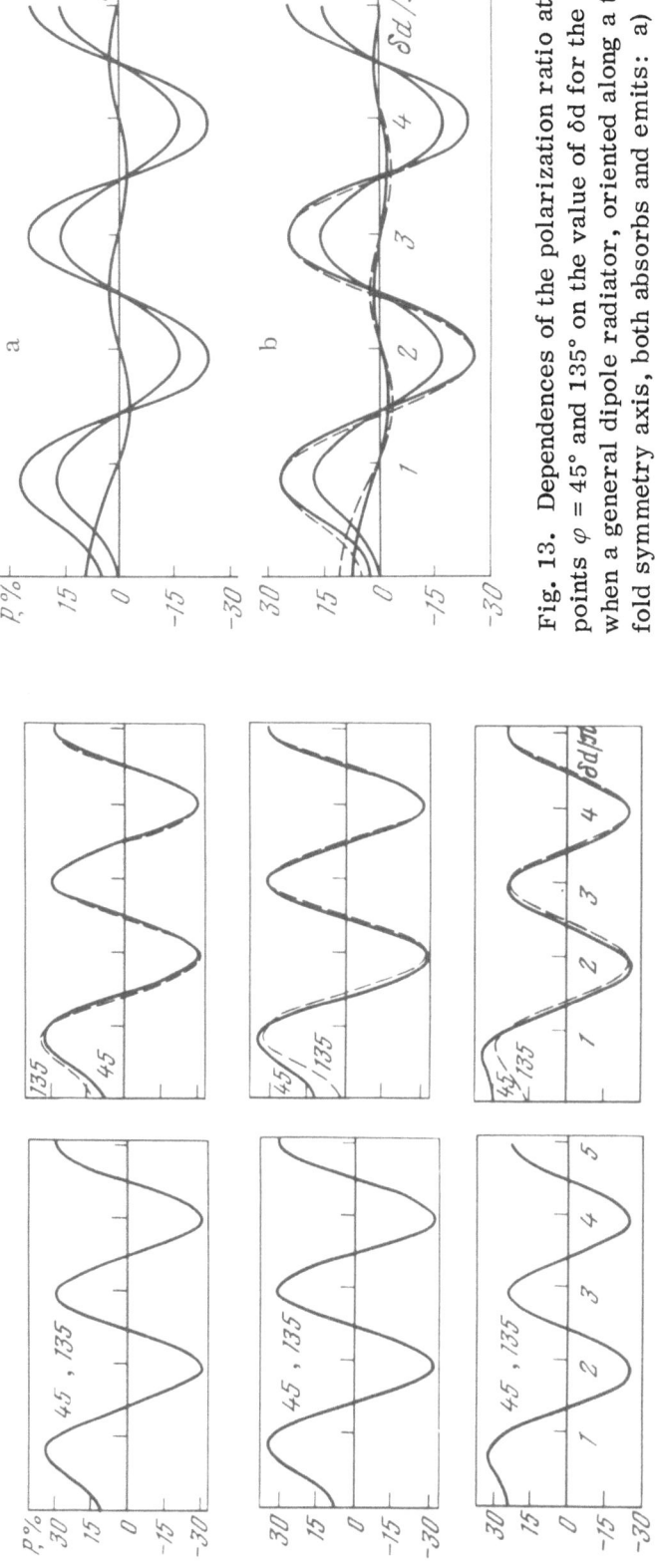

Fig. 13. Dependences of the polarization ratio at the points $\varphi = 45°$ and $135°$ on the value of δd for the case when a general dipole radiator, oriented along a two-fold symmetry axis, both absorbs and emits: a) direction of observation lies in the symmetry plane of the tetrahedron; b) direction of observation perpendicular to this plane. The numbers alongside the curves are the values of the relative contribution β of a rotator. The dashed curves apply to $\varphi = 135°$, unless they coincide with the corresponding curves for $\varphi = 45°$.

Fig. 12. Dependences of the polarization ratio at the points $\varphi = 45°$ and $135°$ on the value of δd for the case when a rotator absorbs and emits. The numbers alongside the curves are the values of φ in degrees. The arrangement is the same as in Fig. 4.

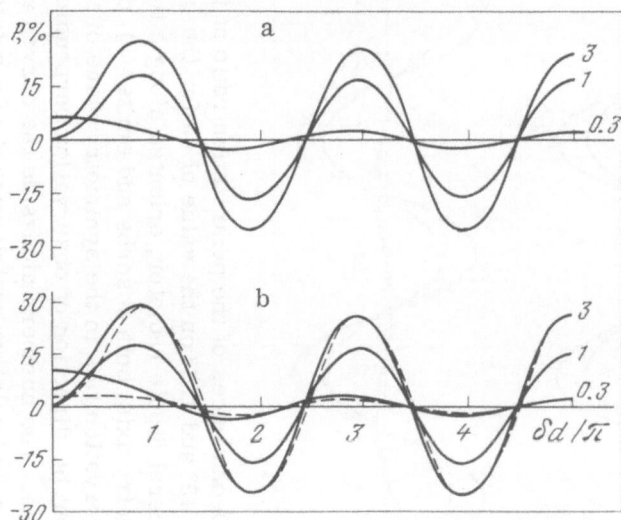

Fig. 14. Dependences of the polarization ratio at
the points $\varphi = 45°$ and $135°$ on the value of δd for
the case when a general dipole radiator, oriented
along a threefold symmetry axis, both absorbs
and emits. The designations are the same as in
Fig. 13.

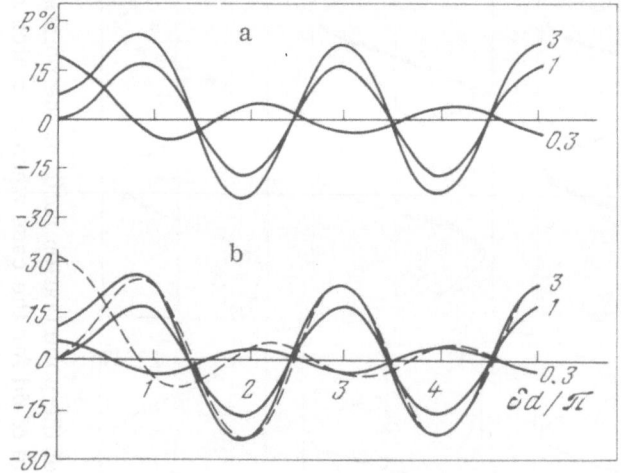

Fig. 15. Dependences of the polarization ratio at
the points $\varphi = 45°$ and $135°$ on the value of δd for
the case when a general dipole radiator, oriented
along a fourfold symmetry axis, both absorbs and
emits. The designations are the same as in
Fig. 13.

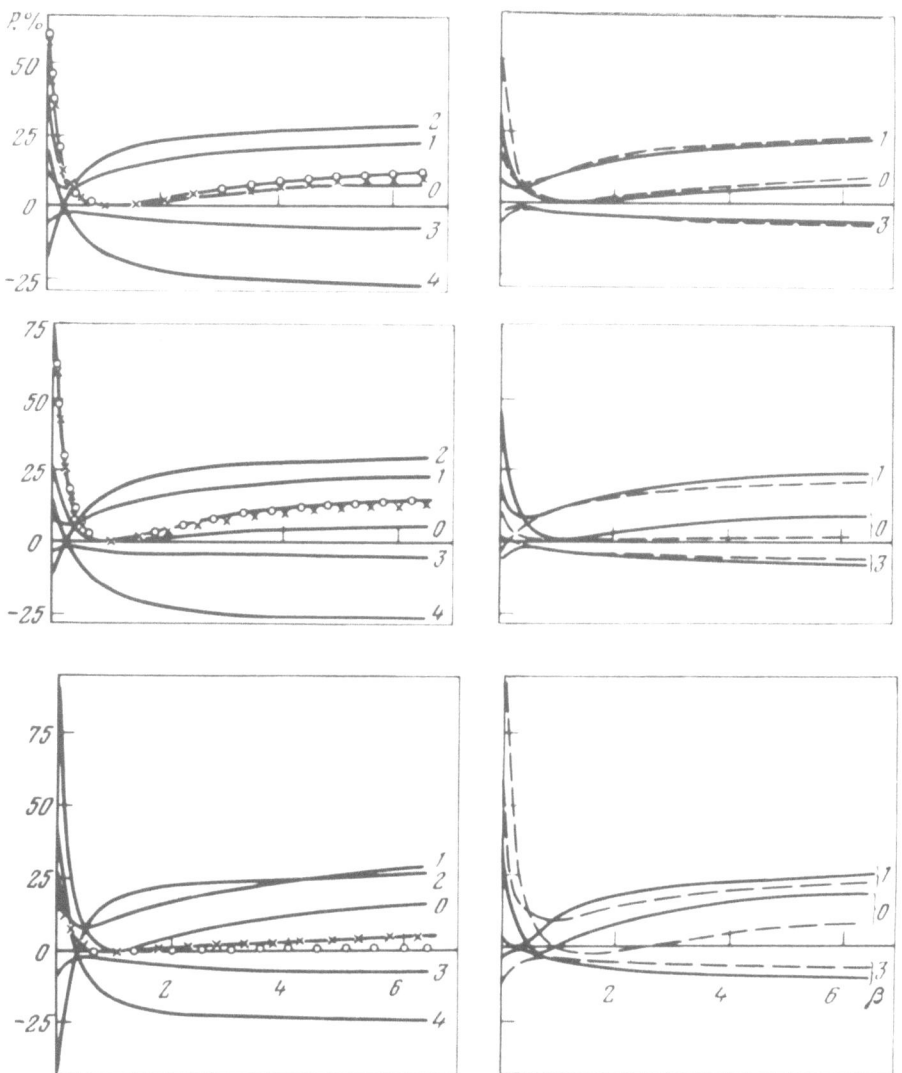

Fig. 16. Dependences, on β, of the polarization ratio at the points $\varphi = 0°$, $45°$, $90°$, and $135°$ for different values of δd. The distribution of the curves is the same as in Fig. 4. The circles correspond to $\varphi = 0°$; the continuous curves correspond to $\varphi = 45°$, the crosses to $\varphi = 90°$, and the dashed curves to $\varphi = 135°$ [if $P(135°) \neq P(45°)$]. The numbers alongside the curves are the values of δd in units of $\pi/2$. The absence of these numbers indicates that the curves in question are independent of δd. The right-hand part of the figure does not include those curves which coincide with the corresponding curves on the left-hand side.

Changes in the polarization diagrams due to a change in δd can easily be followed by examining the figures given in the present section. In some cases a change in δd can even convert a maximum to a minimum and then back to a maximum. The asymmetry of the polarization diagrams is manifested particularly strongly for $\delta d < \pi$, and when δd increases, it becomes smaller. This is readily visible in Figs. 10-12, which show the dependence on δd of P(45°) and P(135°) in the cases corresponding to Figs. 4-6. (We have mentioned earlier that near these points the influence of δd is greatest.) It is clear from these figures that when δd is sufficiently large, the value of P varies almost periodically, and P(45°) is practically equal to P(135°), whereas for $\delta d < \pi$ the differences between them are very large, particularly for radiators oriented along the threefold and fourfold axes.

It is worth considering especially the polarization diagrams for a general dipole radiator ($\beta_e = \beta_l = \beta$). Figures 7-9 and 13-16 give the families of the polarization curves. In all the cases considered above the addition of a rotator ($\beta \neq 0$) to an oscillator reduces strongly the polarization for $\varphi = 0$ and 90° whereas in the region of 45° and 135° we may observe (for some values of δd) an increase in the polarization. The asymmetry also decreases strongly with increasing β ($\beta < 1$). Moreover, it is necessary to point out that the asymmetry decreases more rapidly with rising δd for radiators oriented along the twofold axes than for radiators oriented along the threefold axes.

§ 7. Experimental Investigation of Zinc Sulfide

The nature of the blue luminescence centers of "self-activated" zinc sulfide has not yet been finally established. The doubts persist partly because several different types of center emit in the blue region, and this gives rise to a wide luminescence band which is difficult to analyze into contributions from particular types of center. The generalized Alentsev method was used in [2] to demonstrate that the luminescence of zinc sulfide in the light and dark blue regions was due to centers of at least four different kinds. The profile and position of each of the four individual bands were found in [2]. As a next step in the determination of the nature of these centers we decided to find their orientation in the crystal lattice of zinc sulfide.

We selected zinc sulfide crystals grown by sublimation at 1300°C and not specially activated with any impurity.* These crystals could contain oxygen impurities and, naturally, intrinsic lattice defects. The luminescence spectra of these crystals were investigated in [2].

The polarization diagrams were determined in the spectral regions where, according to [2], the contribution of one of the bands predominated strongly over the other contributions. We found three such regions located near 2.46, 2.72, and 2.90 eV, and in each region we determined the polarization diagrams.† The results obtained for one of the crystals are plotted in Fig. 17.

We shall now see what conclusions can be drawn from the polarization diagrams in Fig. 17. Since the polarization ratio is not equal to zero, it follows that the radiators have a fixed preferred orientation. This means that they are not due to "point" centers, but due to "two-point" centers of the type discussed in the preceding section. The differences between the polarization diagrams determined in different spectral regions confirm the existence of centers of different kinds.

---

* These crystals were kindly supplied by N. A. Gorbacheva, and the authors are grateful to her for this.

† The authors are grateful to Ya. Mokhnyak for the help in this part of the investigation and participation in the discussion of the results obtained.

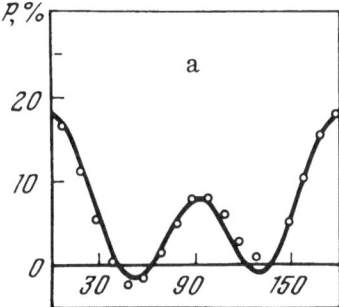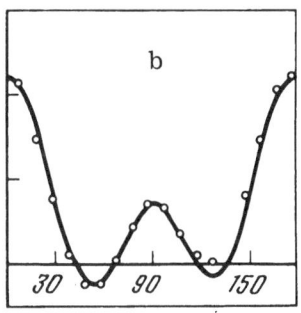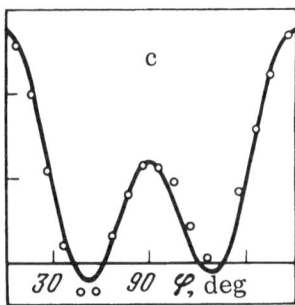

Fig. 17. Comparison of the theoretical (continuous curves) and experimental (points) polarization diagrams for the light blue luminescence of zinc sulfide. The continuous curves represent the theoretical diagrams for the orientation of radiators along the threefold symmetry axes, and the points are the experimental values: a) $h\nu = 2.46$ eV, $\beta = 0.36$, $\varkappa = 1.33$; b) $h\nu = 2.72$ eV, $\beta = 0.34$, $\varkappa = 1.55$; c) $h\nu = 2.90$ eV, $\beta = 0.28$, $\varkappa = 1.48$.

It is clear from Fig. 17 that in all three cases we have $P(0) > P(90°) > 0$, and there are maxima in the polarization diagrams near the 0 and 90° points. Since $P(0)$ does not vanish, it follows that the radiators are oriented along the twofold or threefold, but not the fourfold, axes; since the value of the polarization ratio is positive at these two points, it follows that the same system absorbs and emits because if the linear oscillator absorbs and the rotator emits (or if a rotator absorbs and a linear oscillator emits), we would have had $P < 0$ at the same points.

It is also clear from Fig. 17 that there is a definite asymmetry in the polarization diagrams. This means that the radiators are located asymmetrically relative to the plane defined by the optic axis of the crystal and the direction of observation (ZY plane in our coordinates). The optic axis of the investigated crystals is perpendicular to the (111) plane of the cube, i.e., to the stacking fault plane, and the faults themselves correspond, as mentioned earlier, to rotation of the crystal lattice through 180°. If, as is usually assumed, such rotation occurs about the optic axis, it follows that the radiators should occupy a position symmetric relative to the ZY plane. Since stacking faults are fairly frequent (they occur every few tens of monatomic layers), the distribution of radiators becomes averaged out, and the polarization diagrams should be symmetric. This superposition of two patterns rotated through 180° is observed in the Laue diffraction patterns of these crystals. These patterns have the sixfold (instead of the threefold) symmetry. It follows that the crystal lattice is rotated by each stacking fault layer.

The Laue diffraction patterns and the polarization diagrams recorded for the same crystal can be made to agree if we assume that the lattice is not rotated through 180°, but an inversion takes place at a center at which a sulfur atom is located. This inversion in the sulfur sublattice does not give rise to a stacking fault, whereas in the zinc sublattice all the zinc atoms are in different tetrahedral voids.* This gives rise to an inversion in the positions of reflections in the Laue diffraction pattern. The combination of the direct and inverse Laue patterns gives rise to an apparent sixfold symmetry. The lines joining cation and anion lattice sites remain parallel to the corresponding lines in the noninverted lattice. Therefore, the polarization diagram of a crystal with such a stacking fault remains unaffected, i.e., it remains characteristic of the part of the crystal before a stacking fault. In particular, the asymmetry of the polarization diagram is retained.

---

* In the region of such a stacking fault the distribution of zinc atoms does not correspond to the closest packing.

The maximum values of the polarization ratio found in all three spectral regions did not exceed 30%. This was much less than expected for linear oscillators. On the other hand, special experiments demonstrated that there was hardly any depolarization. The depolarization due to crystal imperfections, deduced by measuring $\cos \delta d$ with the aid of the polarization diagram of light not absorbed in the investigated crystal, was not more than 1-2%. The migrational depolarization was minimized by making the measurements at liquid nitrogen temperature. A study of the afterglow indicated that the polarization ratio was only slightly less than the value obtained during excitation. Thus, all the factors causing depolarization taken together could reduce the polarization ratio by a few percent but not by a factor of 2-3, as found experimentally. Hence, we had to assume that a considerable role was played by rotators in the absorption and luminescence processes, i.e., that we were dealing with general dipole radiators for which the absolute value of $P(\varphi)$ was considerably less than for linear oscillators.

However, in the case of a general dipole radiator the ratios $P(0)$ and $P(90°)$ should decrease by approximately the same amount. It follows from our experimental results that they differ by a factor of more than 2. This difference can be explained by assuming that the equivalent radiators oriented in different ways differ also in respect of their statistical weights. This may happen if the concentration of the investigated luminescence centers in a stacking fault layer is considerably greater than the volume-average concentration because a stacking layer itself gives rise to a preferred direction perpendicular to its plane. An increase in the concentration of the luminescence centers in a stacking fault layer can be explained in the crystal-chemistry sense as follows. Segregation of impurities increases their concentration on the surface which becomes higher than in the bulk of a growing crystal. When the surface concentration of impurities exceeds a certain limit, the growth process is upset and a stacking fault is formed where the accumulated impurities remain; this is followed by normal growth until the process is repeated again. Under these conditions the compensating defects are concentrated in a layer of atoms following the one where the impurities are concentrated and, naturally, these defects are close to the impurities. This results in a preferential orientation of the luminescence centers in a direction close to the stacking fault plane.

Using these ideas we can explain quite satisfactorily the polarization diagrams obtained for the investigated crystals. It is clear from Fig. 17 that the main difference between the experimental and theoretical curves is that the former have deeper minima.

We shall now describe briefly the method for calculation of theoretical polarization diagrams in the case under discussion. The difference between the statistical weights of the radiators $(\varkappa)$ can be allowed for in the summation of $I_{j\parallel}$ and $I_{j\perp}$, some of which differ from unity. Then,

$$P(\varphi) = \frac{\sum_{j=1}^{n} \varkappa_j \left(I_{j\parallel} - I_{j\perp} \right)}{\sum_{j=1}^{n} \varkappa_j \left(I_{j\parallel} + I_{j\perp} \right)}. \tag{47}$$

It follows from our discussion that when radiators are oriented along the twofold axes, we have $\varkappa \neq 1$ for all those radiators whose axes do not lie in the stacking fault plane, i.e., those which have a component parallel to the optic axis. In this case there are three such directions, and all of them make the same angle with the stacking fault plane. Therefore, the value of $\varkappa$ can be assumed to be the same for all these directions.

When radiators are oriented along the threefold axes it is assumed that $\varkappa \neq 1$ only for the radiators whose axes are parallel to the optic axis of the crystal. The other radiators also do not lie in the stacking fault plane, but are inclined at a small (and the same) angle with respect to this plane.

TABLE 5. Polarization Diagram Formulas for Absorption by and Emission from General Dipole Radiators with a Relative Rotator Contribution Equal to β when Some Radiators Have Statistical Weight $\varkappa$ and Others Unity

Radiator orientation	Selected radiators	Direction of observation rel. to tetrahedron symmetry plane	$P(\varphi) = \dfrac{I_{\parallel}(\varphi) - I_{\perp}(\varphi)}{I_{\parallel}(\varphi) + I_{\perp}(\varphi)}$
Parallel to twofold axes	Have component parallel to optic axis of crystal	Parallel	$I_{\parallel} - I_{\perp} = (1 - \beta)\,[\varkappa\,(29 + 43\beta) - 9\,(3 + 5\beta)]\cos 2\varphi + (1 - \beta)^2\,(19\varkappa + 27)\cos^2 2\varphi + 16\,(1 - \beta)^2\,\varkappa \sin^2 2\varphi + 24\beta^2\,(\varkappa + 3)\sin^2 2\varphi\left(\dfrac{\sin \delta d}{\delta d} + \cos \delta d\right),$
		Perpendicular	$I_{\parallel} + I_{\perp} = \varkappa\,(51 + 138\beta + 99\beta^2) + 9\,(3 + 10\beta + 19\beta^2) + (1 - \beta)\,[\varkappa\,(29 + 43\beta) - 9\,(3 + 5\beta)]\cos 2\varphi$
			$I_{\parallel} - I_{\perp} = (1 - \beta)\,[\varkappa\,(29 + 43\beta) - 9\,(3 + 5\beta)]\cos 2\varphi + (1 - \beta)^2\,(19\varkappa + 27)\cos^2 2\varphi - 4\sqrt{2}\,\varkappa\,(1 - \beta)^2\sin 2\varphi\,\dfrac{\sin \delta d}{\delta d} + 4\sqrt{2}\,(1 - \beta)^2\,\varkappa \sin 4\varphi\,\dfrac{\sin \delta d}{\delta d} + 16\varkappa\,(1 - \beta)^2\sin^2 2\varphi + 24\,(\varkappa + 3)\,\beta^2\sin^2 2\varphi\left(\dfrac{\sin \delta d}{\delta d} - \cos \delta d\right),$
			$I_{\parallel} + I_{\perp} = \varkappa\,(51 + 138\beta + 99\beta^2) + 9\,(3 + 10\beta + 19\beta^2) + (1 - \beta)\,[\varkappa\,(29 + 43\beta) - 9\,(3 + 5\beta)]\cos 2\varphi - (1 - \beta)^2\,4\sqrt{2}\,\varkappa\sin 2\varphi\,\dfrac{\sin \delta d}{\delta d}$
Parallel to threefold axes	Parallel to optic axis	Parallel	$I_{\parallel} - I_{\perp} = (1 - \beta)\,[27\varkappa\,(1 + \beta) - (23 + 31\beta)]\cos 2\varphi + (1 - \beta)^2\,(27\varkappa + 17)\cos^2 2\varphi + 8\,(1 - \beta)^2\sin^2 2\varphi\left(\dfrac{\sin \delta d}{\delta d} + \cos \delta d\right) + 72\beta^2\sin^2 2\varphi\left(\dfrac{\sin \delta d}{\delta d} - \cos \delta d\right),$
		Perpendicular	$I_{\parallel} + I_{\perp} = 27\varkappa\,(1 + \beta)^2 + 3\,(11 + 38\beta + 59\beta^2) + (1 - \beta)\,[27\varkappa\,(1 + \beta) - (23 + 31\beta)]\cos 2\varphi$
			$I_{\parallel} - I_{\perp} = (1 - \beta)\,[27\varkappa\,(1 + \beta) - (23 + 31\beta)]\cos 2\varphi + (1 - \beta)^2\,(27\varkappa + 17)\cos^2 2\varphi + 8\sqrt{2}\,(1 - \beta)^2\sin 2\varphi\,\dfrac{\sin \delta d}{\delta d} - 8\sqrt{2}\,(1 - \beta)^2\sin 4\varphi\,\dfrac{\sin \delta d}{\delta d} + 8\,(1 - \beta)^2\sin^2 2\varphi\left(\dfrac{\sin \delta d}{\delta d} + \cos \delta d\right) + 72\beta^2\sin^2 2\varphi\left(\dfrac{\sin \delta d}{\delta d} - \cos \delta d\right),$

TABLE 5. Continued

Radiator orientation	Selected radiation	Direction of observation rel. to tetrahedron symmetry plane	$P(\varphi) = \dfrac{I_\parallel(\varphi) - I_\perp(\varphi)}{I_\parallel(\varphi) + I_\perp(\varphi)}$
Parallel to three-fold axes	Parallel to optic axis	Perpendicular	$I_\parallel + I_\perp = 27\varkappa(1+\beta)^2 + 3(11+38\beta+59\beta^2) + (1-\beta)[27\varkappa(1+\beta)-(23+31\beta)]\cos 2\varphi + 8\sqrt{2}(1-\beta)^2 \sin 2\varphi \dfrac{\sin\delta d}{\delta d}$
Parallel to fourfold axes	Lie in plane formed by optic axis and direction of observation	Parallel	$I_\parallel - I_\perp = (1-\beta)[2\varkappa(1+5\beta)-5-7\beta]\cos 2\varphi + (1-\beta)^2(2\varkappa+1)\cos^2 2\varphi + 12(1-\beta)^2\sin^2 2\varphi\left(\dfrac{\sin\delta d}{\delta d}+\cos\delta d\right) + 12\beta^2\sin^2 2\varphi\left(\dfrac{\sin\delta d}{\delta d}-\cos\delta d\right)(2\varkappa+1),$ $I_\parallel + I_\perp = 2\varkappa(1+5\beta)^2 + (5+7\beta)^2 + (1-\beta)[2\varkappa(1+5\beta)-5-7\beta]\cos 2\varphi$
	Perpendicular to direction of observation	Perpendicular	$I_\parallel - I_\perp = 3(1-\beta)[1+3\beta-2\varkappa(1+\beta)]\cos 2\varphi + (1-\beta)^2(2\varkappa+1)\cos^2 2\varphi + 6\sqrt{2}(1-\beta)[2\varkappa(1+\beta)-1-3\beta]\sin 2\varphi\dfrac{\sin\delta d}{\delta d} - 2\sqrt{2}(1-\beta)(2\varkappa+1)\times$ $\times \sin 4\varphi \dfrac{\sin\delta d}{\delta d} + 4(1-\beta)^2(2\varkappa+1)\sin^2 2\varphi\left(\dfrac{\sin\delta d}{\delta d}+\cos\delta d\right) + 36\beta^2\sin^2 2\varphi\left(\dfrac{\sin\delta d}{\delta d}-\cos\delta d\right).$ $I_\parallel + I_\perp = 18\varkappa(1+\beta)^2 + 9(1+3\beta)^2 + 3(1-\beta)[1+3\beta-2\varkappa(1+\beta)]\cos 2\varphi + 6\sqrt{2}(1-\beta)[2\varkappa(1+\beta)-(1+3\beta)]\sin 2\varphi\dfrac{\sin\delta d}{\delta d}$

When radiators are oriented along the fourfold axes, all three directions of the radiators form the same angle with the stacking fault plane, and there are no grounds for preferring any one direction. Moreover, as mentioned earlier, in this case the polarization diagrams do not correspond to those found experimentally. Therefore, there is no need to calculate the polarization diagrams for this orientation. However, for the sake of generality, we can derive the formulas for the polarization diagrams for this case, as well, on the assumption that the statistical weight of one of the types of radiator differs from the statistical weights of the other two. The formulas for all these cases are given in Table 5.

The values of $\varkappa$ and β which agreed best with the experimental results were found by solving a system of two equations of the type $P_{theor}(0) = P_{exp}(0)$ and $P_{theor}(90°) = P_{exp}(90°)$. The values of $\varkappa$ and β found in this way ensured that the theoretical curves passed through these two experimental points. At other points the theoretical and experimental values could differ. In particular, it is clear from Fig. 9 that if the parameter β was selected incorrectly, minima and not maxima would be found between these points. In the calculation of the polarization diagrams we had to find also the value of δd. This quantity was found in two stages. The value of $\cos \delta d$ was deduced from the shape of the polarization diagram for nonabsorbed light, as described in § 1. (For the crystal whose polarization diagrams are plotted in Fig. 17 we found that $\cos \delta d = 0.05$.) The value of δd was estimated by analogy with other crystals of the same type for which our data yielded $\delta = 1000-1100$ deg/mm. Since the crystal thickness was ~1 mm, the value of $\cos \delta d$ agreed with δd if the latter was assumed to be 993° or 17.3 rad. The next root of this trigonometric equation was outside the range of likely values of δ for the investigated crystals.

We found that the polarization diagrams calculated subject to these assumptions for radiators oriented along the twofold and threefold axes were almost identical except that the minima in the latter case were somewhat deeper (these are the curves plotted in Fig. 17). The values of $\varkappa$ were found to be greater than unity for all three photon energies, and these values were quite close to one another. Moreover, the values of β were also close, and they were less than unity. This indicated a relatively small rotator contribution, which nevertheless altered strongly the polarization diagrams. The low value of the polarization ratio of cubic crystals was probably due to the same reason.

When a crystal was rotated by 90° around the optic axis, the polarization diagrams became (as expected) much more symmetric. In particular, the maxima of $P(\varphi)$ were found to be located exactly at $\varphi = 0$ and 90°. However, some asymmetry in the region of the minima was still observed. It was probably due to an insufficiently accurate orientation of the crystal when a sample of the necessary shape was cut from it. The orientation error was 2-3°, which could affect significantly the value of $P(\varphi)$ in the region of $\varphi = 45°$ and 135°. Nevertheless, the polarization diagram method enabled us to determine several important details of the structure of the luminescence centers found in the investigated birefringent crystals. The present authors hope that other investigators will adopt this method.

Literature Cited

1. P. P. Feofilov, The Physical Basis of Polarized Emission: Polarized Luminescence of Atoms, Molecules, and Crystals, Consultants Bureau, New York (1961).
2. E. E. Bukke, T. I. Voznesenskaya, N. P. Golubeva, N. A. Gorbacheva, Z. P. Kaeeva, E. I. Panasyuk, and M. V. Fok, Tr. Fiz. Inst. Akad. Nauk SSSR, 59:25 (1972).

LUMINESCENCE POLARIZATION INVESTIGATION OF THE STRUCTURE OF DOPED MOLECULAR CRYSTALS

N. D. Zhevandrov and T. V. Il'inykh

An investigation was made of the spatial distribution of the polarization of the luminescence emitted by naphthalene and anthracene single crystals with one or two impurities, and also of the dependence of the polarization of the impurity luminescence on the polarization of the exciting radiation. This investigation yielded information on the orientation of the impurity molecules and on the nature of their incorporation in the host lattice. It was found that impurity molecules did not show a tendency to form pairs. The sensitivity of this method for investigating the structure of doped molecular crystals was five orders of magnitude higher than the sensitivity of the x-ray structure method.

Introduction of molecules, which normally form a molecular crystal, into the lattice of a different molecular crystal alters considerably the luminescence intensity of these molecules [1, 2]. In some cases, which are particularly interesting from the theoretical and practical points of view, the intensity rises strongly. This is closely related to the nature of the incorporation of the impurity molecules into the host lattice, orientation of these molecules, and deformation of such molecules and of the immediate environment, as well as other structure factors. On the other hand, the processes involving the transfer of energy from the host molecules to the impurities and between the impurities also depend strongly on the position and orientation of the impurity molecules. Therefore, the results of structural studies of the incorporation and orientation of impurity molecules in the host lattice are of considerable interest.

Unfortunately, the x-ray structure analysis method is not sufficiently sensitive and quite unsuitable when the impurity concentration is low. We shall tackle this task by the luminescence polarization method based on an investigation of the spatial distribution of the polarization emitted from oriented single crystals (polarization diagram method).

The method of luminescence polarization diagrams was developed by the present authors earlier and used to determine the orientation of luminescence-emitting oscillators relative to the molecular axes, and also the orientation of the molecule itself in the host lattice and of the impurity molecules added to the host crystal in concentrations of 0.1-0.01% (measured in the original charge) [3-8].

We applied this method to the structure of several doped crystals based on naphthalene and anthracene, and we attempted to select host−impurity pairs to be able to study doped crystals with low impurity concentrations.

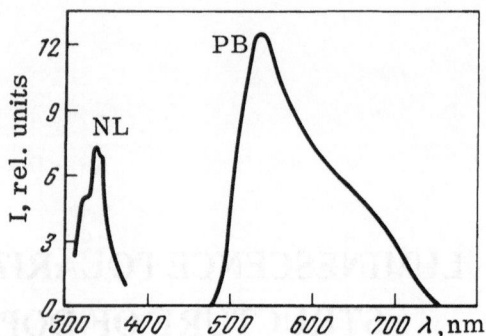

Fig. 1. Luminescence spectrum of a mixed naphthalene crystal (NL) containing phosphor B (PB) in a concentration of 10^{-5} g/g.

The results demonstrated that the luminescence polarization method could be used to determine not only the orientation of molecules in the lattice or oscillators in the molecule, but, as indicated by the preliminary analysis, to tackle a much wider range of structural tasks.

§ 1. Structure Investigations of Doped Naphthalene Single Crystals

We investigated the orientation of the molecules of phosphors A and B, naphthacene, anthracene, and anthranilic acid in the crystal lattice of naphthalene. These materials were selected because of their luminescence spectra and because of the variety of molecular structures, which should give rise to different types of incorporation of impurities into the host lattice. This should be manifested clearly in the investigated polarization diagrams.

Figure 1 shows the luminescence spectrum of a mixed crystal of naphthalene containing phosphor B. Clearly, the luminescence spectra of naphthalene and phosphor B did not overlap. This was also true of the spectra of naphthalene containing phosphor A and naphthacene.

We also found that the luminescence efficiency of the phosphors A and B increased strongly when they were incorporated in naphthalene. The fortunate location of the impurity luminescence spectra (in the yellow-green region, where the human eye has the maximum sensitivity) and the bright luminescence of the impurities incorporated in the host lattice allowed us to investigate crystals with very low impurity concentrations (right down to 10^{-7} g/g), to obtain for these crystals the luminescence polarization diagrams, and to deduce from them the orientation of the impurity molecules in the host lattice.

Naphthalene crystals were grown in sealed glass ampoules on oriented seeds in vertical two-section tubular heaters through which ampoules were lowered slowly.*

Naphthalene crystals were grown using a sublimated reagent of "analytic purity" grade, which was additionally purified by the growth of single crystals from the melt on an oriented single-crystal seed. The extreme (most strongly contaminated) parts were removed from such crystals. The middle part of a crystal was used to grow a new single crystal. We found that crystals obtained by this double-crystallization method contained no more than 10^{-7} g/g foreign impurities capable of luminescence in the 450-700-nm range, which was fully satisfactory for our purposes. Therefore, doped crystals were grown from a reagent which was purified once by growing a single crystal from the melt. The second purification occurred directly during the growth of a doped crystal in the presence of luminescent admixtures.

---

* These single crystals were grown at the Institute of Crystallography, Academy of Sciences of the USSR, by Candidates of Chemical Sciences G. S. Belikova and L. E. Kraeva. The present authors are very grateful to them for this service.

The phosphor A and B impurities in naphthalene crystals were not expelled from the bulk of the single crystal and neither did they precipitate as microcrystals; instead, they formed a uniformly colored and a uniformly luminescent single crystal. The polarization studies confirmed the formation of a homogeneous doped single crystal and not a heterophase system.

Our crystals contained impurities in concentrations from 10^{-7} to 10^{-3} g/g. When impurities were present in the melt in concentrations of 10^{-2} g/g or higher, a single crystal was not obtained. In this case strong stresses were observed in the lattice, and these caused cracking.

The impurity concentration in a crystal was determined by the spectroscopic method of comparing standard solutions of known concentration with solutions prepared directly from lumps of the investigated single crystals.

Detailed data on the crystal structure of naphthalene were taken from [10]. The distribution of molecules in a unit cell is shown in Fig. 2. Naphthalene crystals belong to the monoclinic system of the Fedorov class $P2_1/a$, and a unit cell contains two molecules oriented in different ways and located at the lattice sites and at the centers of the (001) faces. The unit cell dimensions are $a = 8.235$ Å, $b = 6.003$ Å, $c = 8.658$ Å, and $\beta = 122°55'$. The orientation of the molecules at the lattice sites is given by the angles between the molecular and crystallographic axes:

	L	M	N
a	115°.8	7°.2	32°.8
b	102.8	29.45	116.3
c'	29.0	68.2	71.9

Here, L, M, and N are the axes of the naphthalene molecule (L is the longitudinal axis, M is the transverse axis, and N is the normal to the plane of the molecule); a, b, and c' are the crystal axes (a and b are the crystallographic axes, and c' is the axis perpendicular to the ab plane).

The positions of the molecules on the centers of the (001) faces are derived from the positions of the molecules at the lattice sites by reflection in the (010) plane.

Naphthalene is a biaxial crystal exhibiting a strong birefringence. The optic axes plane (010) coincides with the symmetry plane of the crystal. The monoclinic structure of naphthalene

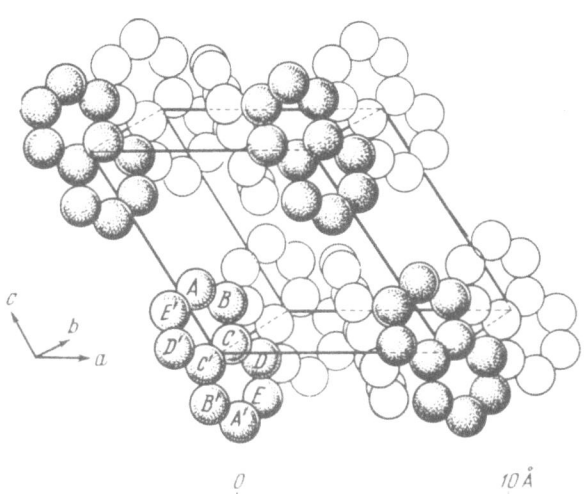

Fig. 2. Distribution of molecules in a unit cell of naphthalene.

is characterized by a close packing of the molecules in the *ab* layer; repetition of these layers by translation along the *c* axis is less dense [10], which explains the perfect cleavage of naphthalene along the (001) planes. The polarization diagrams, i.e., the dependences of the degree of polarization of the emitted luminescence on the direction of observation for a given orientation of a crystal, were measured using the visual polarization goniometric system.

Impurities were excited by natural light either directly (light from a PRK-4 mercury lamp of wavelength 436 nm, which was located outside the absorption band of the host crystal) or via the host (light of wavelength 313 nm). In the latter case the excitation energy migrated from the host to the impurity. The luminescence spectra of the impurities were shifted quite considerably relative to the spectrum of the host substance, and they could be selected conveniently with a ZhZS-9 filter. The degree of polarization was measured with an improved Cornu polarimeter based on the equalization of two illumination fields which were produced by a Wollaston prism and represented the mutually perpendicular polarized components. This polarimeter differed from the usual Cornu instrument by a circular entry aperture and the presence, behind the Wollaston prism, of a Fresnel biprism, which brought together the two mutually perpendicular polarized images of the entry aperture using half of each image. A sharp photometric boundary between these fields was always formed by the biprism edge.

Time-consuming calculations of the corrections for the luminescence birefringence were avoided by the use of hemispherical samples excited at the center of the sphere on the flat side. The luminescence polarization was measured along the radial directions of the sphere as a function of the angle of observation χ in the horizontal plane. Thus, the luminescence emerging from a hemispherical sample was always directed normally to the surface. This minimized the error associated with the luminescence birefringence. A hemispherical crystal was oriented in the desired way by selecting the plane of cut of a sphere and then rotating the resultant hemisphere about a horizontal axis [4]. A large single crystal grown in a test tube was sawed with a wire into parts of the required size, and spheres were formed in a rotating tube; they were then polished using the same tube but covered with chamois leather. Then, the resultant transparent spheres were oriented. The cleavage planes were deduced from a series of cracks which appeared when a sample was cooled by wetting with a rapidly evaporating substance (such as ether or acetone). The positions of the optic axes were determined by the Shubnikov method [11]. A conoscopic figure of a biaxial crystal demonstrated clearly the point where a given optic axis emerged. Since the angle between the optic axes in naphthalene crystals was large, the point of emergence of one axis could be found first, and then the point of emergence of the second axis could be established by rotating this sphere.

Spheres were reduced to hemispherical shape by grinding along the following crystallographic planes: 1) along the optical axes plane (010), denoted by *ac*; 2) along the cleavage plane (001), denoted by *ab*; 3) along a plane perpendicular to the *ac* and *ab* planes, i.e., along the *bc'* plane.

A direct calculation of all the angles between the molecular and crystallographic axes on the basis of the experimentally determined polarization diagrams would be very complex and would give ambiguous results. Therefore, we calculated the polarization diagrams using the x-ray structure data on the host lattice and on the orientation of the luminescence oscillators in the impurity molecules, as well as the available information on the orientation of the impurity molecules. A comparison of the experimental and calculated diagrams then enabled us to accept or reject a given hypothesis.

In our case it was possible to assume that the impurity molecules replaced the host molecules in such a way that the planes of the benzene rings and the longitudinal axes of the naphthalene and impurity molecules were nearly parallel. When the directions of the luminescence oscillators in the impurity molecules were also known, all the diagrams of interest to us could be calculated. It was known that luminescence oscillators in anthracene molecules were directed along the transverse axis [4]. Moreover, it was likely that the luminescence oscillators in the phosphor A and B molecules were directed along the longitudinal axis.

Figure 3 shows the polarization diagrams of naphthalene calculated on the assumption that the luminescence oscillators were directed along the longitudinal and transverse axes; this was done for all the cuts mentioned above [7]. These cuts were selected because they gave the most specific diagrams in respect of the orientation of the crystal and of the lumines-

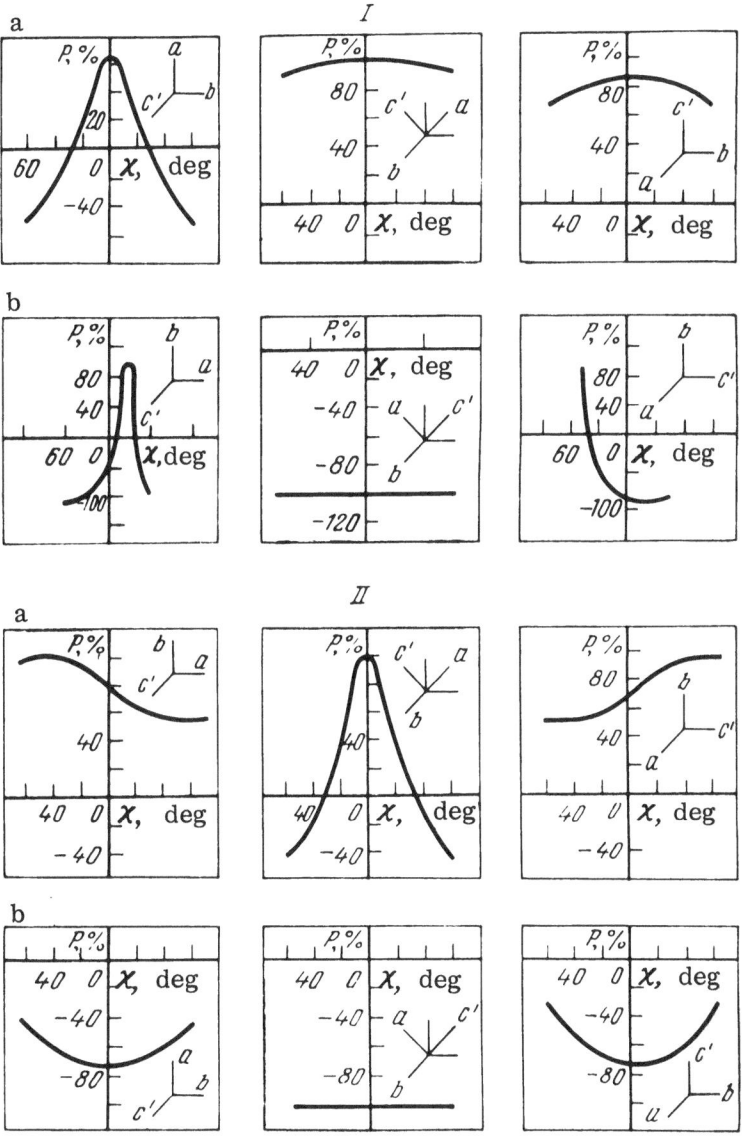

Fig. 3. Calculated positive (a) and negative (b) polarization diagrams of naphthalene for oscillators oriented along the longitudinal (I) and transverse (II) molecular axes. Here, χ is the angle of observation measured in the horizontal plane to the right or left of the normal; P is the degree of polarization of the luminescence. The orientation of a crystal is shown alongside each curve.

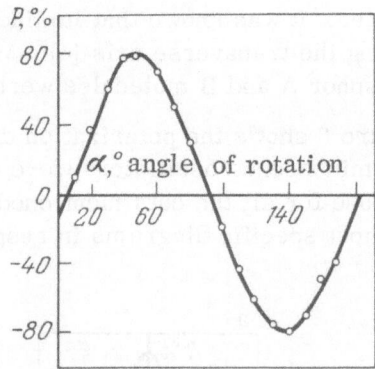

Fig. 4. Azimuthal dependence of the degree of
polarization.

cence oscillators.   This made it possible to compare the calculated and experimental polarization diagrams and then determine unambiguously the orientation of the impurity molecules
in the crystal lattice of naphthalene.

Since the degree of polarization was independent not only of the angle of observation, but
also of the orientation of the crystal itself, we determined the azimuthal dependence of the
polarization for each of the cuts, i.e., we found the dependence of the degree of polarization on
the angle of rotation of a crystal about an axis coinciding with the optic axis of the polarimeter
(Fig. 4).

The orientations exhibiting the extremal polarizations were used to determine the "positive" and "negative" polarization diagrams in the horizontal plane.

The experimentally determined positive and negative polarization diagrams of the phosphors A and B in naphthalene were plotted (Figs. 5-8).   The calculated and experimental dia-

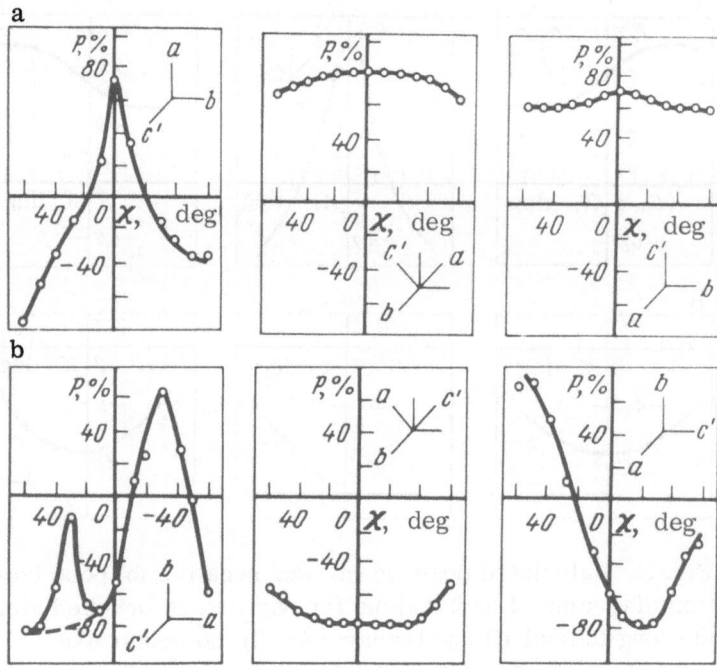

Fig. 5. Experimental polarization diagrams of phosphor
A in naphthalene (λ_{exc} = 313 nm).   The notation is the
same as in Fig. 3.

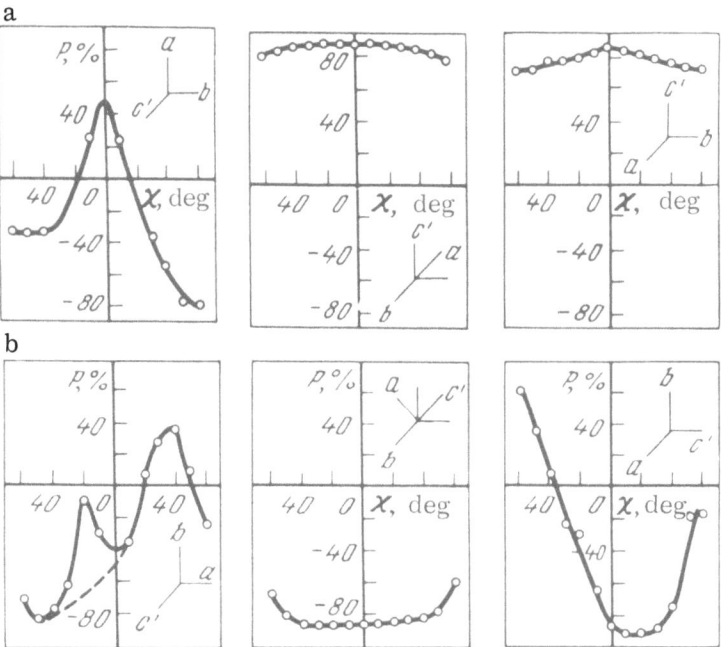

Fig. 6. Experimental polarization diagrams of phosphor B in naphthalene (λ_{exc} = 313 nm). The notation is the same as in Fig. 3.

grams were in agreement if it was assumed that the oscillators were oriented along the longitudinal molecular axis for both phosphors A and B. This confirmed our hypothesis that the introduction of impurities into the naphthalene lattice occurred in such a way that the longitudinal axis of the impurity molecules became parallel to the longitudinal axis of naphthalene. A comparison of the geometries of the phospor A and the naphthalene molecules, carried out by finding the projections of the models of these molecules plotted using the intermolecular radii [10, 12], demonstrated that the projections of the models of the phosphor A molecules could be made to coincide with two projections of the naphthalene models. It was likely that

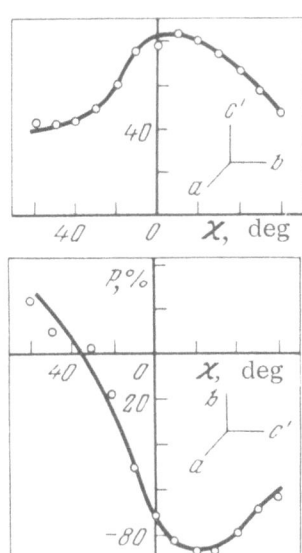

Fig. 7. Experimental polarization diagrams of phosphor A in naphthalene (λ_{exc} = 436 nm, C = 10^{-5} g/g).

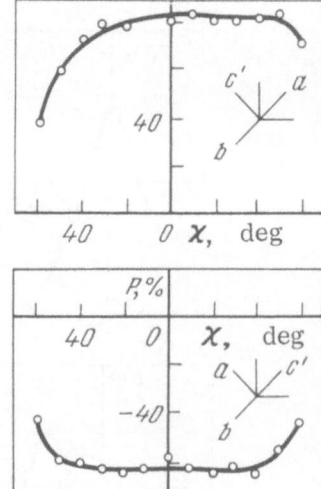

Fig. 8. Experimental polarization diagrams of
phosphor B in naphthalene (C = 10^{-7} g/g).

the phosphor A molecules were incorporated in the naphthalene lattice in such a way that one impurity molecule replaced two neighboring naphthalene molecules oriented parallel along the c axis (this was the direction of the least packing of the molecules).

Thus, the replacement process produced a substitutional and not an interstitial solid solution. This conclusion was supported by an interesting observation that the investigated naphthalene single crystals containing the two phosphors (A and B) were mechanically stronger than pure naphthalene (this was observed clearly in the preparation of single-crystal samples), and the mechanical strength increased with the impurity concentration. The impurity molecules acted as "bracing links" along the c axis, which corresponded to the largest lattice constant.

The incorporation of the impurity molecules in the host lattice "at the two ends" undoubtedly strengthened and stabilized the structure of naphthalene, reduced the possibility of vibrations, and, clearly, was responsible for the strong increase in the luminescence efficiency of the impurity molecules compared with the efficiency of the impurities in bulk.

The correctness of our conclusions on the nature of the incorporation and orientation of the phosphor A and B molecules in naphthalene was confirmed by a strong dichroism. Only light with a nonzero projection of the radiation electric vector onto the direction of the absorption oscillator was absorbed in the crystals. If this light traveled along the oscillator direction, the transverse nature of the light waves would ensure that they would not be absorbed at all, and the crystal should be uncolored along this direction. Since the long-wavelength absorption oscillators in the A and B phosphor molecules were oriented parallel to the long axis and the impurity molecules were incorporated into the host lattice in such a way that their long axes were almost parallel to the c axis, a crystal viewed along the c axis should be uncolored, which was confirmed by direct visual observations.

At first sight it would seem that the polarization diagram method should yield only the orientation of molecules in the lattice. However, our results indicated that when additional data were used, one could tackle also a wider range of structural tasks.

It should be noted that the calculated diagrams were located further from the abscissa than the experimental ones. One of the causes of this quantitative discrepancy was the thermal depolarization, but the depolarizing influence of the rotation and vibration of molecules about their equilibrium positions in molecular crystals was considerably less than in solutions and was a negligible quantity [4]. Clearly, the main factor was that the luminescence polarization diagrams were calculated using the linear oscillator model, i.e., purely electronic transitions were considered. Allowance for the electron-vibrational (vibronic) transitions led to the presence of not just one linear oscillator but of a cone of oscillators. The replacement of such a cone with one effective oscillator overestimated the degree of polarization.

The best agreement between the experimental and calculated diagrams was obtained for the samples cut in the optical axes plane; the agreement was poorer for the samples cut in the cleavage plane. This was clearly due to the fact that naphthalene single crystals split easily along the cleavage plane and because rays with the same polarization suffered different reflection losses by microcracks formed on these planes; this distorted the true degree of the luminescence polarization. Therefore, it was necessary to examine the samples closely so as to ensure that the spheres obtained were free of all possible defects.

Since naphthalene sublimated readily in air even at room temperature, all the measurements were carried out by placing a hemisphere in a holder inside a cylindrical fused quartz cell filled with distilled water. This prevented sublimation and minimized depolarization because of the scattering by the surface of a crystal. In measurements of the polarization diagrams a hemisphere was placed inside the cylindrical cell in such a way that the direction corresponding to one of the polarization maxima was parallel to the generators of the cell. For this orientation of the hemisphere relative to the cell the measurement plane coincided with the circular cross section of the cylinder. This minimized the depolarization due to the refraction in the cell walls.

The distortion of the measured degree of polarization because of the birefringence in the crystal itself was reduced by the use of small crystals and by ensuring the greatest possible overlap of two split beams. The small difference between the experimental and calculated diagrams indicated that the replacement of two naphthalene molecules with one impurity molecule produced slight local stresses in the lattice. However, these stresses were negligible because the impurity molecules occupied spatially favorable positions in the host lattice.

The experimentally determined polarization diagrams obtained for the samples cut along the cleavage plane had marked "dips" in the angular range $\chi = 20\text{-}30\%$. The positions of these dips corresponded to the points of emergence of the optic axes. This was not accidental. The dips were clearly due to the internal conical refraction. In this phenomenon one wave normal, coinciding with an optic axis, corresponded to a cone of rays which emerged from a crystal to form a cylindrical beam of rays of different polarizations. The overall degree of polarization of the whole beam was not be affected, and the ratio of the two polarization components did not change. However, these components became spatially separated so that when the polarimeter was moved in the plane of observation, its entry aperture received first the part of the beam with one polarization and then with another, so that the measured degree of polarization differed from the true value. These distortions affected particularly the samples cut along the cleavage plane ab. This was because in the case of the samples cut in the bc' plane the point of emergence of an optic axis coincided with a maximum in the polarization diagram, so that the distortions caused by the internal conical refraction simply increased the maximum or lowered it, which did not produce such large effects as the dips in a monotonic curve.

The presence of these dips did not complicate greatly the comparison of the experimental and calculated diagrams because interpolation could easily be made across the dips.

We mentioned earlier that the impurity luminescence was excited either via the host lattice ($\lambda_{exc} = 313$ nm − Figs. 5 and 6) or directly ($\lambda_{exc} = 436$ nm − Fig. 7). The experimental diagrams obtained in these two cases were in equally good agreement with the calculated diagrams, which was to be expected because the luminescence polarization should be independent of the method of energy transfer to the molecules emitting luminescence. Then we obtained the luminescence polarization diagrams for crystals with different impurity concentrations. The measurements were carried out on naphthalene single crystals which contained A and B phosphors in concentrations of 10^{-3}-10^{-7} g/g. We found that the diagrams obtained were independent of the impurity concentration in mixed molecular crystals. This indicated that, irrespective of the concentration, the impurities were incorporated in the same way in the host lattice. Hence, the distribution of the molecules in this lattice should be randomly independent and there should be no pair formation.

The bright luminescence and the fortunate lack of overlap of the luminescence spectra enabled us to determine the polarization diagrams down to impurity concentrations of 10^{-7} g/g. Figure 8 shows, by way of example, the polarization diagrams of a naphthalene single crystal with this concentration of phosphor B. The diagrams were found to agree well with the calculations. However, in the case of naphthalene crystals grown without any additive we were unable to obtain any regular polarization diagrams in the green part of the spectrum. Hence, we concluded that an undoped naphthalene single crystal, purified by double crystallization, contained less than 10^{-7} g/g impurities emitting visible luminescence.

All the results demonstrated a high sensitivity of the polarization diagram method, compared with the x-ray structure method, whose sensitivity in the case of mixed crystals was limited to impurity concentrations exceeding 2-3%. The polarization diagram method enabled us to determine the orientation of the impurity molecules in the host lattice down to concentrations of 0.00001%, i.e., the sensitivity was improved by five or six orders of magnitude.

We also investigated the nature of the incorporation and orientation of anthracene molecules in the naphthalene lattice. Figure 9 shows the experimentally determined polarization diagrams of anthracene in naphthalene. Once again the excitation was provided via the host lattice by light of wavelength λ = 313 nm. The luminescence spectrum of anthracene was selected by an SS-14 filter.

A comparison of the experimental results (Fig. 9) with calculations (Fig. 3) demonstrated that the experimental diagrams were in good agreement with those calculated for oscillators directed along the transverse axis. The luminescence oscillator in the anthracene molecules was known to be oriented along the transverse axis of these molecules. Consequently, an-

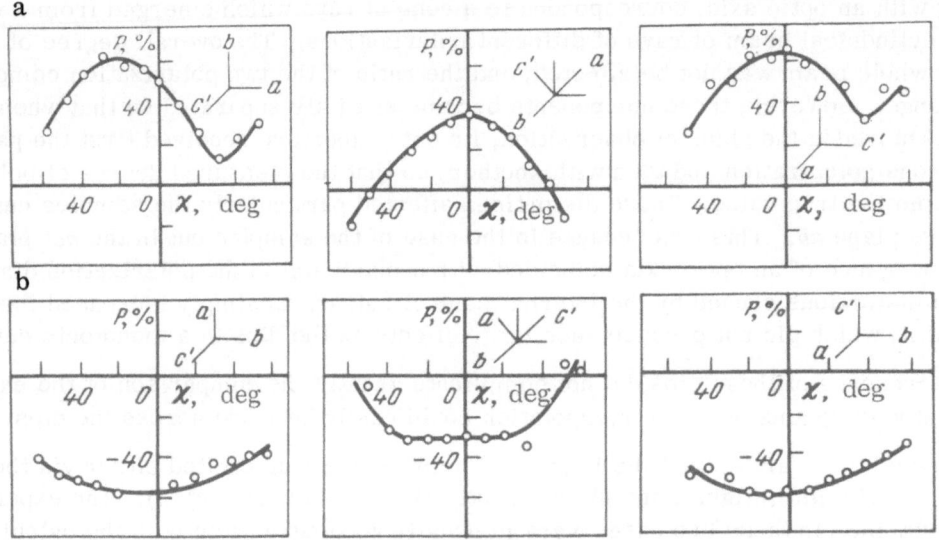

Fig. 9. Experimental polarization diagrams of anthracene in naphthalene
(λ_{exc} = 313 nm).

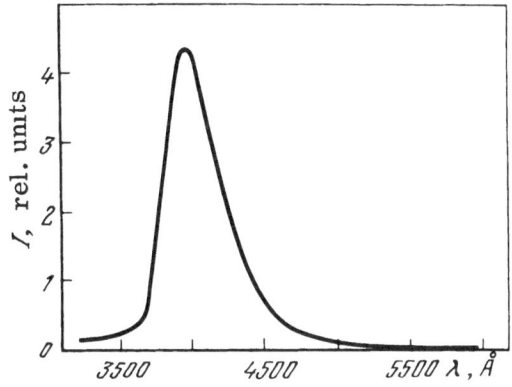

Fig. 10. Luminescence spectrum of a naphthalene single crystal containing anthranilic acid ($C = 10^{-4}$ g/g, $\lambda_{exc} = 313$ nm).

thracene molecules replaced isomorphously the naphthalene host molecules. However, this replacement was difficult to achieve because each anthracene molecule differed by one benzene ring from the naphthalene molecule (the length of the naphthalene molecule along the longitudinal axis L was ~7 Å and that of anthracene was ~10 Å), which resulted in considerable local stresses in the host lattice. This was demonstrated by some distortion of the experimental diagrams, compared with the calculated curves, and by the limited amount of anthracene that could be incorporated in naphthalene during crystal growth (up to $C = 10^{-4}$ g/g). The distortions in the crystal lattice were supported by qualitative observations of the appearance of a second cleavage plane oriented at 60° with respect to the first one.

Our aim was to study incorporation of impurity molecules of different structures. The phosphor A and B molecules each replaced two naphthalene host molecules. Each anthracene molecule replaced one naphthalene molecule but this gave rise to local stresses. For comparison, we decided to study an impurity whose molecules would be as close as possible to naphthalene in respect of their shape, dimensions, and oscillator orientation. In this case we could expect nearly isomorphous substitution. It was interesting to confirm this experimentally by our method.

These requirements were found to be satisfied by the anthranilic acid. We investigated mixed crystals of naphthalene and anthranilic acid containing 10^{-3}-10^{-5} g/g of the impurity. The luminescence spectrum of anthranilic acid in naphthalene single crystals was recorded (Fig. 10). The luminescence spectrum of the host was located at shorter ultraviolet wavelengths. The impurity luminescence was selected with an SS-8 filter.

Figure 11 shows the experimentally obtained polarization diagrams of anthranilic acid in naphthalene. The experimental diagrams were in good agreement with those calculated for oscillators parallel to the transverse axis of the molecule (i.e., oscillators oriented in the same way as naphthalene itself). This confirmed the assumption that anthranilic acid molecules were incorporated in the host lattice in such a way that the corresponding molecular axes of the impurities and naphthalene and of the planes of their benzene rings were nearly parallel. A comparison of the geometry of the naphthalene and anthranilic acid molecules, made using the intermolecular radii, indicated that the projection of an anthranilic acid molecule fitted well the projection of a naphthalene molecule. Consequently, we concluded that these impurities were incorporated in such a way that one anthranilic acid molecule replaced one host molecule.

We investigated also naphthalene single crystals containing naphthacene. The naphthacene molecule was two benzene rings longer than the naphthalene molecule. The luminescence oscillator was directed along the transverse (short) axis of the naphthacene molecule [13-15]. We used single crystals with the optimal naphthacene concentrations of 10^{-4} and 10^{-5} g/g. At

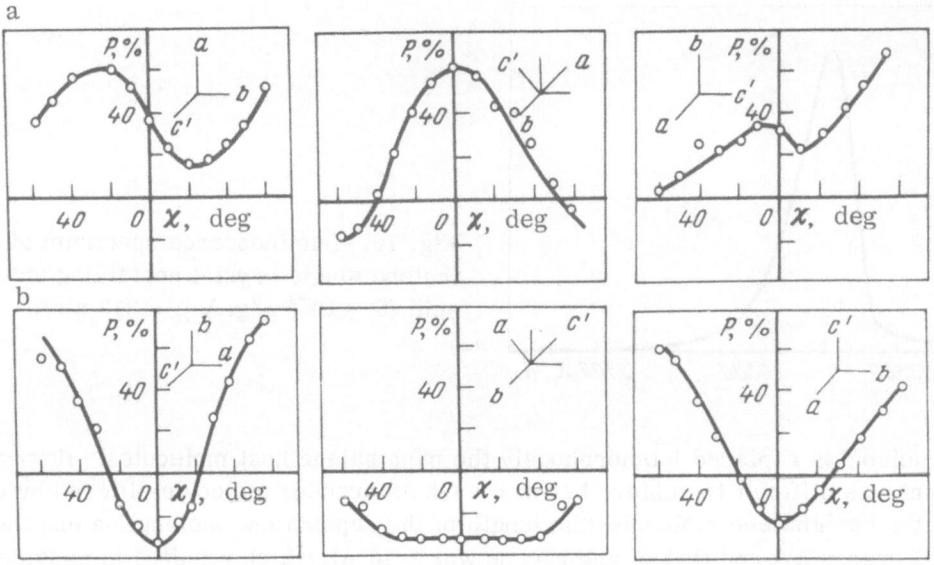

Fig. 11. Experimental polarization diagrams of anthranilic acid in naphthalene (λ_{exc} = 313 nm).

these concentrations naphthacene was readily incorporated in the naphthalene lattice and it emitted bright luminescence.

We analyzed the experimentally determined polarization diagrams of naphthacene in naphthalene (Fig. 12). They agreed well with the diagrams calculated for oscillators oriented along the longitudinal axis of the naphthalene molecule, i.e., they differed considerably from the diagrams of anthracene in naphthalene. Consequently, the naphthacene and anthracene molecules were incorporated in different ways in the naphthalene host lattice. The following conclusion was drawn from the transverse orientation of the naphthacene oscillator and from the structure of the crystal lattice of naphthalene and the molecular structure of naphthacene. The naphthacene molecules were incorporated in the crystal lattice of naphthalene in such a

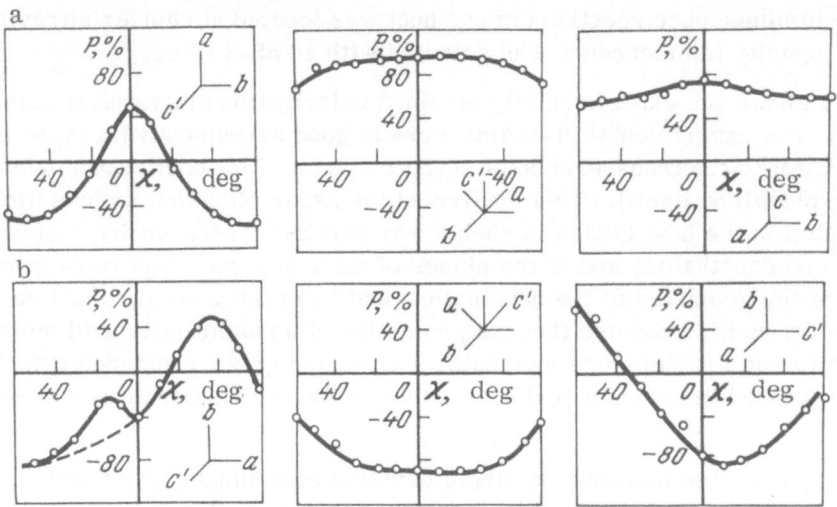

Fig. 12. Experimental polarization diagrams of naphthacene in
naphthalene (λ_{exc} = 313 nm, C = 10^{-4} g/g).

way that one naphthacene molecule replaced two naphthalene molecules parallel to the b axis; the planes of the benzene rings of the naphthalene and naphthacene molecules were parallel and the longitudinal axis of the naphthacene molecule coincided with the transverse axis of the naphthalene molecule.

The different orientations and method of incorporation of various impurity molecules (naphthacene, anthracene, phosphors A and B) made it desirable to study single crystals containing two different impurities in order to determine the influence of one impurity on the nature of incorporation and orientation of molecules of the other impurity [16, 17].

§2. Structure Investigations of Crystals
with Two Impurities

Since the transfer of energy from the host lattice to various impurities and from molecules of one impurity (donor) to another (acceptor) depends strongly on the orientation and relative positions of the impurity molecules in the host lattice, investigations of the structure of molecular crystals with two impurities are highly desirable.

In experimental studies of the migration of energy in doped crystals it is very important to know how the impurity molecules are incorporated in the host lattice, i.e., whether they are distributed in pairs or singly and independently of one another. Some of the experimental results given below support the hypothesis that impurity molecules are distributed singly in a statistically independent (random) manner.

We used the luminescence polarization diagram method to determine whether molecules of two different impurities were incorporated in pairs or singly in the host lattice. We studied naphthalene single crystals containing phosphor A and anthracene as impurities, as well as naphthalene single crystals containing anthracene and naphthacene.

We grew and investigated single crystals with the same concentrations of both impurities (10^{-4} and 10^{-5} g/g), as well as crystals with different impurity concentrations: 10^{-5} g/g of anthracene and 10^{-3} g/g of phosphor A; $5 \cdot 10^{-4}$ g/g of anthracene and 10^{-6} g/g of phosphor A.

When high concentrations of the phosphor A, anthracene, or naphthacene were used, single crystals of poor quality were obtained beginning from 10^{-3} g/g impurity concentration, and this was due to the appearance of strong local stresses in the lattice causing cracking.

The process of measurement of the polarization diagrams provided means for the control of the quality and purity of the grown single crystals. Regular polarization diagrams were not obtained for crystals of poor quality. Moreover, the polarization diagram method made it possible to check the absence of other impurities exhibiting visible luminescence right down to concentrations of 10^{-7} g/g.

The selected substances were found to be very suitable for the diagram polarization method for the following reasons. Firstly, the use of this method required accurate selection (with suitable filters) of the luminescence first of one and then of the other impurity. This could be done readily in the investigated cases because the luminescence spectra of naphthalene, anthracene, and phosphor A had hardly any overlap (Fig. 13). The luminescence spectrum of the phosphor A was selected by a ZhZS-9 filter and that of anthracene, by an SS-14 filter. This was also true of the spectra of anthracene and naphthacene in naphthalene.

Secondly, the luminescence oscillator in the phosphor A molecule was directed along the longitudinal axis of this molecule, whereas in anthracene it was directed along the transverse axis, so that it was possible to obtain basically different polarization diagrams for the differently oriented oscillators of the two impurities in the same crystal. This basic difference between

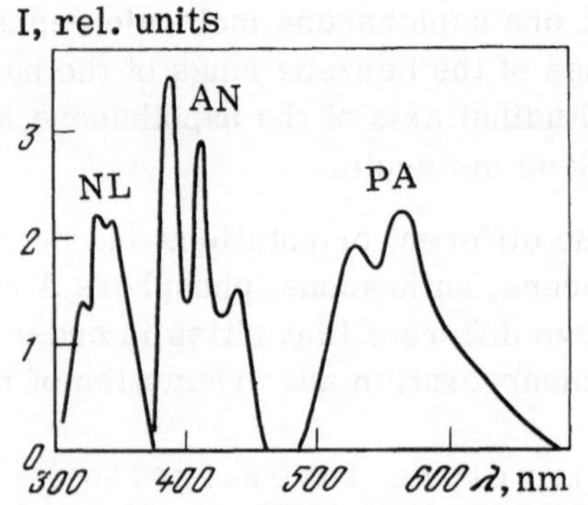

Fig. 13. Luminescence spectrum of a two-impurity naphthalene crystal NL doped with anthracene AN and phosphor PA.

the diagrams of two different impurities in the same host lattice, each of which agreed well with the corresponding calculations, increased considerably the reliability of the results obtained.

In the case of naphthacene the luminescence oscillator was directed along the transverse axis of the molecule (exactly in the same way as in anthracene). However, the nature of the incorporation of anthracene in naphthalene was quite different from the incorporation of naphthacene in naphthalene. Therefore, it was of considerable interest to investigate the polarization diagrams of these two impurities in the same naphthalene host single crystal.

Positive and negative polarization diagrams were obtained experimentally for naphthalene crystals containing both anthracene and phosphor A (Fig. 14). The orientation of a crystal was indicated in each case alongside the diagram. In the case of cuts along the ab and bc' planes the orientation of a crystal with the maximum positive polarization of the phosphor A luminescence was the same as that for the maximum negative polarization of the anthracene luminescence; in the case of samples cut along the ac plane the positions with the maximum positive and negative values of the degree of polarization of anthracene were the same as for phosphor A, in agreement with the calculations. Hence, it was possible to select the luminescence of each of the two impurities in a crystal not only by spectral, but also by polarization, methods. Using crystal cuts for which the polarization diagrams were antibatic, we could remove the luminescence of the second impurity by a suitably oriented Polaroid.

A comparison of the experimental and calculated results indicated that the polarization diagrams for phosphor A agreed well, as in the case of a crystal with just one impurity, with the diagrams calculated on the assumption that the oscillators were directed along the longitudinal axis (compare diagram I in Fig. 3 with diagram I in Fig. 14); in the case of anthracene, the experimental results were compared with the diagrams calculated for the oscillators directed along the transverse axis (diagram II in Fig. 3 and diagram II in Fig. 14). The dip in the experimental diagrams in Fig. 14 (II) obtained for a sample cut in the ab plane indicated the direction of the optic axis and was due to internal conical refraction; it was easy to interpolate across this dip (dashed curve) [9].

Consequently, the phosphor A and anthracene impurities were incorporated in the naphthalene lattice in such a way that the longitudinal axes and the planes of the benzene rings of the naphthalene and impurity molecules were nearly parallel. A comparison of the molecular dimensions indicated that one phosphor A molecule replaced two naphthalene molecules parallel to the c axis, whereas one anthracene molecule replaced one naphthalene molecule. The polarization diagrams were very similar to those obtained for the corresponding one-impurity crystals.

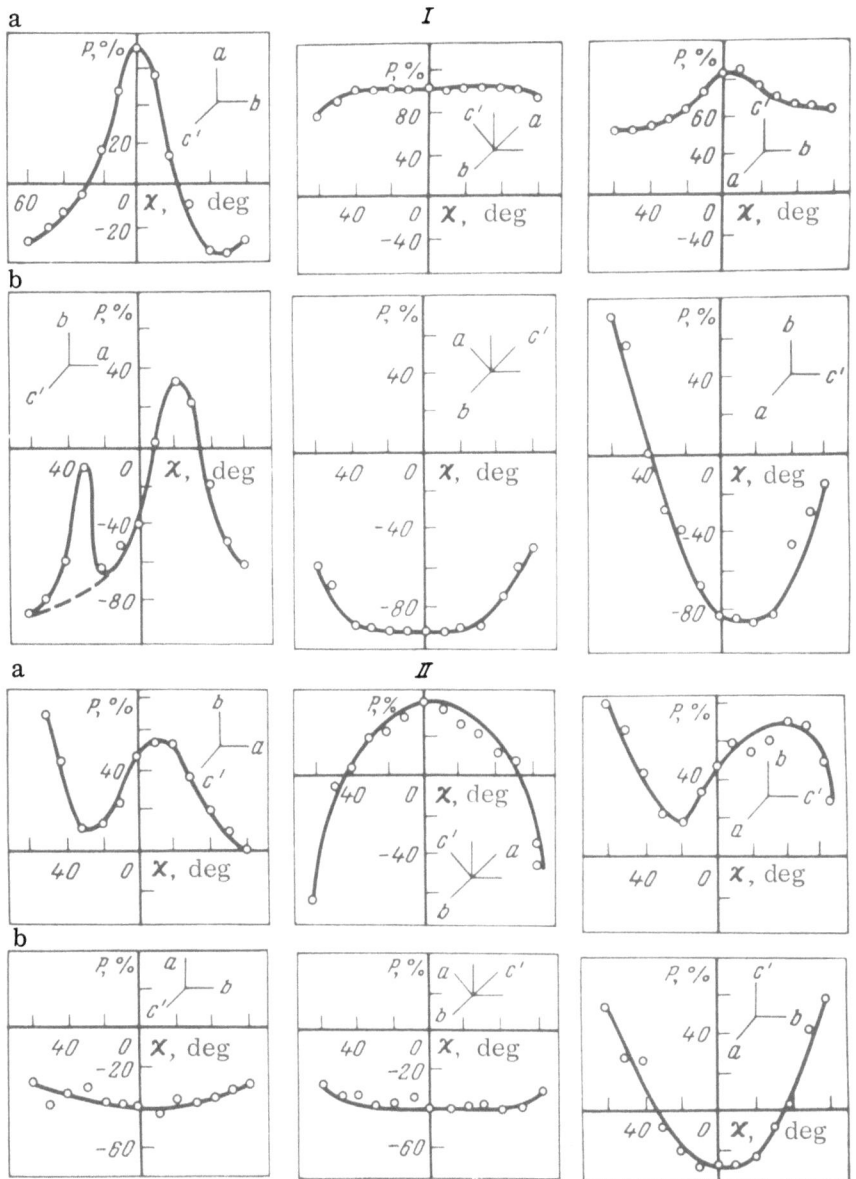

Fig. 14. Experimental polarization diagrams of the phosphor A (I)
and anthracene (II) in naphthalene.

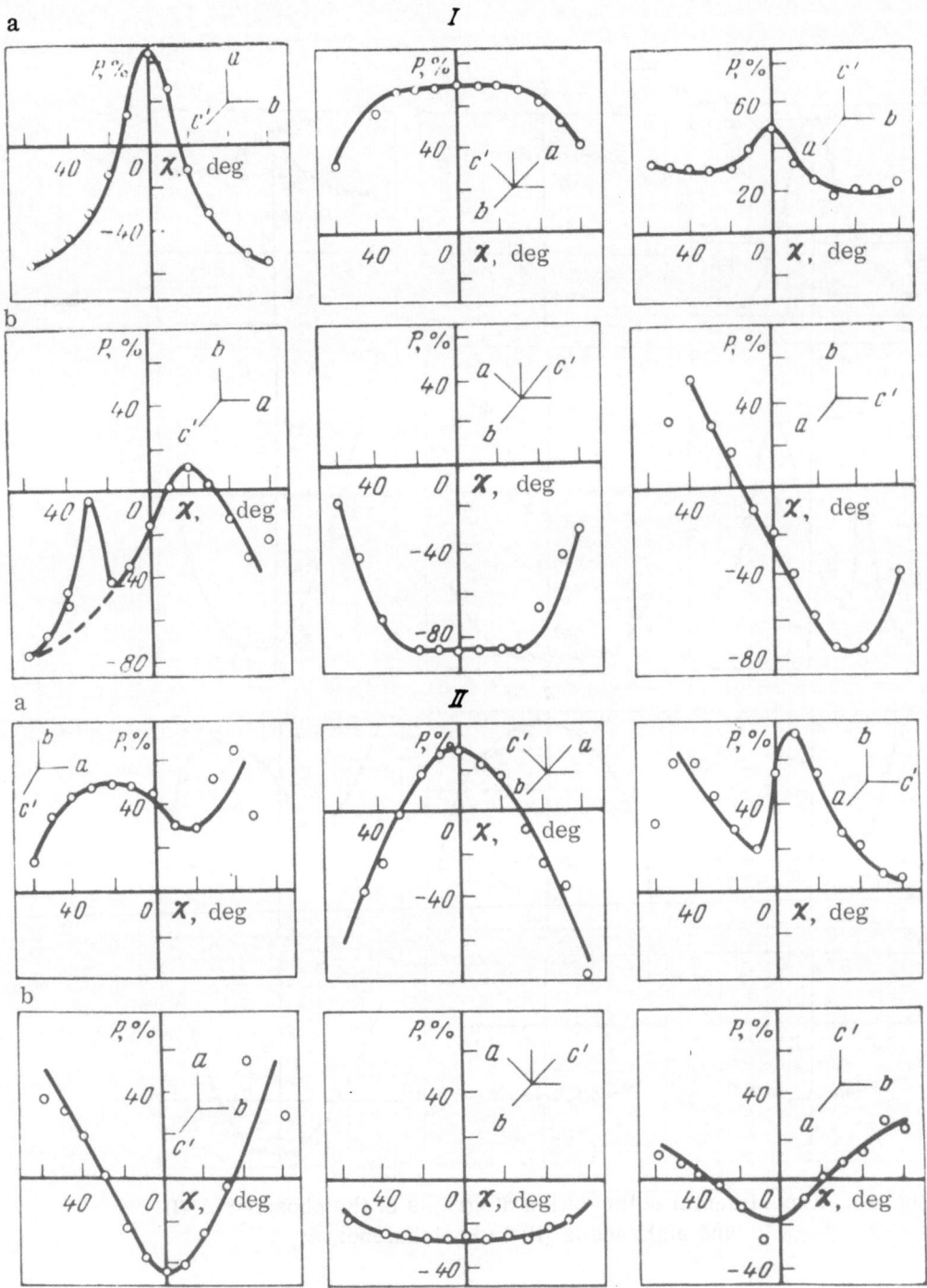

Fig. 15. Experimental polarization diagrams of naphthacene (I) and anthracene in naphthalene ($C_{NC} = C_{AN} = 10^{-4}$ g/g).

Figure 15 shows the experimentally obtained polarization diagrams of naphthalene containing naphthacene and anthracene impurities. A comparison of these diagrams with the corresponding one-impurity results (Figs. 9 and 12) indicated that there were no significant changes in the diagrams of the two-impurity crystals compared with those obtained for one-impurity samples.

Thus, in the investigated two-impurity molecular crystals the impurity molecules were oriented in the host lattice exactly in the same way as in the corresponding one-impurity crystals. The good agreement between the polarization diagrams of the two-impurity crystals with the corresponding diagrams of the one-impurity samples indicated that the presence of one impurity did not affect the nature of incorporation and orientation of the other impurity because the presence of the second impurity did not alter significantly the polarization diagrams of the first impurity.

This result indicated that the molecules of two different impurities incorporated in the same naphthalene host lattice did not have any tendency toward formation of pairs, i.e., there was no tendency toward replacement of neighboring host molecules in the same or nearby unit cells. Such a tendency would have resulted in local stresses induced by molecules of one impurity, and this would have distorted significantly the polarization diagrams of the second impurity.

We determined the polarization diagrams of two-impurity crystals with identical and different concentrations of the two impurities (10^{-5} g/g of anthracene and 10^{-4} g/g of phosphor A; $5 \cdot 10^{-4}$ g/g of anthracene and 10^{-6} g/g of phosphor A). The good agreement between the experimental diagrams and the calculated results indicated that the nature of incorporation and orientation of one impurity was independent of the concentration of the other impurity.

The experimental results indicated that molecules of different impurities were incorporated in the naphthalene lattice in a random manner exhibiting no tendency for the formation of pairs.

§ 3. Structure Investigations of Doped
Anthracene Single Crystals

Difficulties were encountered in the purification of the original reagent used to grow anthracene single crystals. Therefore, these crystals were obtained from the highest-purity synthetic anthracene, purified additionally to remove mechanical and chemical impurities through sublimation in sealed evacuated glass ampoules. Similar ampoules were used to grow single crystals of pure anthracene and of anthracene with deliberately introduced admixtures [17].

We grew single crystals of anthracene with 10^{-4} and 10^{-5} g/g of naphthacene, as well as crystals containing phosphor 11 or phosphor 12 in concentrations of 10^{-4} g/g. Since phosphor A dissolved with difficulty in the anthracene lattice, poor-quality single crystals were obtained, and it was not possible to carry out structure investigations on these crystals by the polarization diagram method. The crystal lattice parameters of anthracene were known from x-ray structure studies [10]. Anthracene crystals, like those of naphthalene, were known to be monoclinic. A unit cell contained two differently oriented molecules located at the lattice sites and at the centers of the (001) faces. The dimensions of a unit cell were $a = 8.56$ Å, $b = 6.04$ Å, $c = 11.16$ Å, and $\beta = 124°42'$. The orientation of the molecules at the lattice sites could be represented by the angles between the molecular and crystallographic axes:

	L	M	N
a	119°.7	71°.3	36°.2
b	97	26.6	115.5
c'	30.6	71.8	66.2

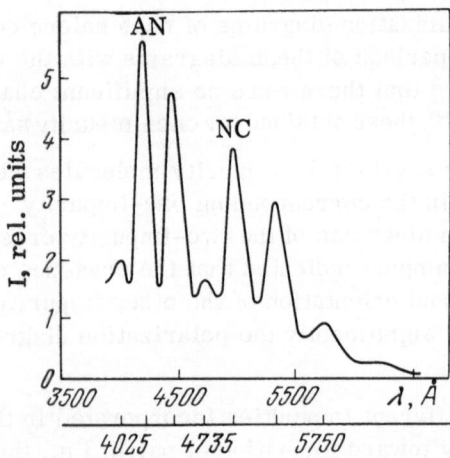

Fig. 16. Luminescence spectrum of a single crystal of anthracene AN doped with naphthacene NC.

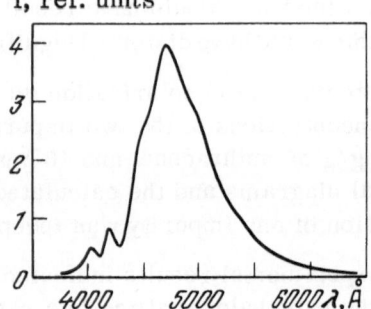

Fig. 17. Luminescene spectrum of a single crystal of anthracene doped with phosphor 11.

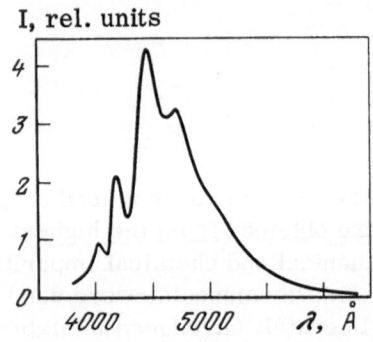

Fig. 18. Luminescence spectrum of a single crystal of anthracene doped with phosphor 12.

The polarization diagrams of doped anthracene single crystals were recorded without recourse to water immersion; this was in contrast to the technique used to obtain the diagrams of doped naphthalene single crystals. The impurity was excited through the host lattice by light of wavelength 365 nm.

Figure 16 shows the luminescence spectrum of naphthacene-doped anthracene, Fig. 17 gives the spectrum of anthracene doped with phosphor 11, and Fig. 18 the spectrum of anthracene doped with phosphor 12. It is clear from these figures that the impurity luminescence could be selected reliably by filters only in the case of naphthacene-doped anthracene. The luminescence spectra of the phosphors 11 and 12 overlapped strongly the luminescence spectrum of the anthracene host and this gave rise to very serious experimental difficulties.

The calculated polarization diagrams of anthracene were practically identical with those given in Fig. 3 for naphthalene with corresponding cuts, which was due to the similarity of the structures of these two compounds.

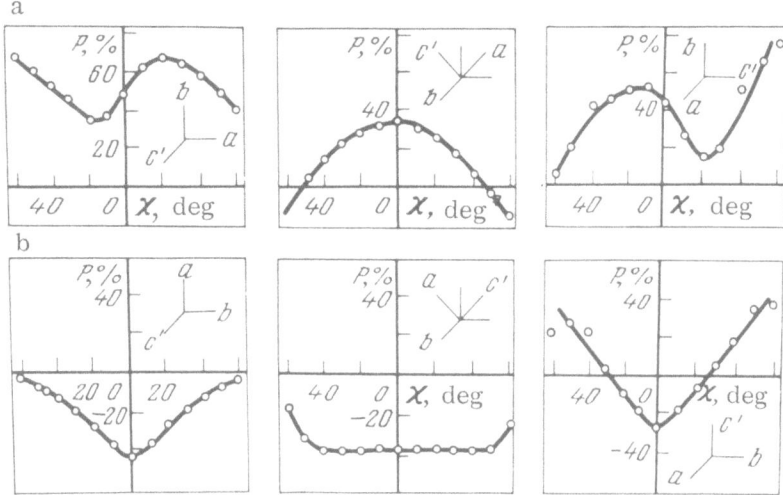

Fig. 19. Experimental polarization diagrams of naphthacene
in anthracene.

Figure 19 shows the experimentally determined polarization diagrams of naphthacene-doped anthracene. These diagrams were in good agreement with the calculations when it was assumed that the oscillator was directed along the transverse axis of the molecule. Since the luminescence oscillator in the naphthacene molecule was directed along the transverse axis of the molecule, we concluded that the naphthacene molecule replaced isomorphously the anthracene molecules in the lattice, i.e., one naphthacene molecule replaced one anthracene molecule in such a way that the planes of the benzene rings and the long axes of the impurity and host molecules were identical. However, the difference between the host and impurity molecules (amounting to one benzene ring) resulted in some local stresses in the lattice, which distorted the diagrams somewhat.

It was interesting to study also anthracene crystals with impurities whose molecules could replace two anthracene molecules parallel to the c axis. The results obtained for naphthalene crystals doped with phosphors A and B (described earlier) indicated that the incorporation of such impurities in the anthracene lattice should result in an increase of the mechanical strength and a considerable increase in the luminescence efficiency compared with the impurity substance in bulk. An impurity molecule should have a structure and dimensions to make this substitution possible. It was difficult to select suitable impurities because the melting point of anthracene was fairly high ($T_{mp} = 216.5°C$), and, therefore, it was necessary to select impurity molecules which would be stable under the prolonged action of high temperatures and, moreover, would be sufficiently long to replace two anthracene molecules. We tried various phosphors and found that phosphors 11 and 12 were most suitable: These were incorporated in the anthracene lattice during single-crystal growth and they had more or less suitable molecular lengths.

We grew anthracene single crystals containing either phosphor 11 or phosphor 12 in a concentration of 10^{-4} g/g. However, it was very difficult to separate the luminescence spectra of these impurities from the luminescence spectrum of anthracene (Figs. 17 and 18). We used a combination of two filters, which were OS-12 and ZhZS-9. The most characteristic polarization diagrams for the longitudinal and transverse luminescence oscillators were obtained for samples cut in the cleavage plane (ab). The positive and negative polarization diagrams obtained experimentally for this cut were strongly distorted by the influence of the host luminescence on the luminescence of the two phosphors. Thus, it was necessary to

determine experimentally the "mixed" diagrams of the whole luminescence emitted by the
impurity crystals. A comparison with the known diagrams of pure anthracene made it possible to establish the qualitative nature of the impurity diagrams. This did not give results as
reliable as in the direct measurements of the impurity diagrams. However, the results indicated qualitatively that the impurity diagrams corresponded to the longitudinal oscillator case.
Hence, we concluded that the molecules of phosphors 11 and 12 were incorporated in the anthracene lattice in such a way that one impurity molecule replaced two host molecules parallel
to the c axis; the long axes and planes of the benzene rings of impurity and host molecules
were all parallel.

§ 4. Relationship between the Structure of Doped Crystals and Energy Migration between Impurity Molecules

In the preceding sections we demonstrated that the polarization diagrams can give
valuable information on the structure of doped molecular crystals. Such information is not
confined only to the orientation data. We shall now give the results of investigations carried
out using a different polarization method also capable of giving information on important details of the structure of such crystals. We used the luminescence polarization to study the
transfer of energy between identical impurity molecules in the host lattice [18].

Experimental studies of the transfer of energy between different molecules can be carried out quite easily because of the wide range of the available methods based on the sensitization phenomenon and related to the difference between the molecules themselves.

The situation is different in the case of energy transfer between identical molecules of
a given substance. In this case the luminescence polarization methods are the most suitable.
A typical example is the concentration depolarization in liquid solutions.

A fine method for investigating energy migration between molecules in molecular crystals is based on the dependence of the luminescence polarization on the polarization of the
exciting light. This method is particularly valuable in studies of the energy transfer in doped
crystals between identical impurity molecules.

The principle of this luminescence polarization method is described in detail in [4]. It
can be summarized briefly as follows. It follows from the principle of close packing that a
molecular crystal has at least two molecular oscillators with different orientations. Since
they are generally oriented in different ways with respect to the radiation electric vector of
linearly polarized exciting light, the probabilities of excitation of these oscillators may be different. Therefore, the degree of polarization of the light emitted by oscillators of different
orientation can be different. Consequently, if there is no energy transfer between such molecules, the degree of luminescence polarization should depend strongly on the orientation of the
radiation electric vector of the exciting light relative to the luminescence oscillators, i.e.,
relative to the crystal axes. However, if such migration does occur during the lifetime of the
excited state, the initial anisotropy is affected. Consequently, the luminescence polarization
will be independent of the orientation of the radiation electric vector of the exciting light. The
polarization will always be the same, as in the case of isotropic excitation with natural light.
Thus, the absence or presence of this dependence provides a sensitive means for the detection
of energy migration between molecules in the molecular crystal lattice.

It would be particularly interesting to apply this method to independently excited impurity
molecules.

We studied naphthalene single crystals containing phosphor admixtures. Incorporation of these phosphor molecules in the naphthalene lattice occurred (as reported above) by the substitution of one impurity molecule for two neighboring naphthalene molecules oriented parallel to the c axis; the orientation of the impurity molecules was identical or very close to the orientation of the naphthalene molecules in the host lattice. The agreement between the calculated and experimental polarization diagrams indicated that the probability of the replacement of naphthalene molecules of different orientations (at the lattice sites and on the centers of the faces) was the same.

In the first approximation, all the impurity molecules could be described using the oriented gas model.

In measurements of the luminescence polarization as a function of the polarization of the exciting light a very important aspect was the orientation of the crystal relative to the axes along which the mutually perpendicular components were polarized in the measuring system; these components governed the degree of the luminescence polarization. For example, when a crystal was oriented in such a way that the bisector of the angle between the oscillator axes coincided with the direction of one of the polarimeter axes, the predicted dependence should not be observed even in the absence of migration because the ratio of the projections onto the polarimeter axes for each of the oscillators should be the same and the degree of polarization should be independent of the orientation of the radiation electric vector of the exciting light and for unpolarized excitation.

The most favorable was the orientation of a single crystal in which one of the oscillators (I or II) coincided with one of the polarimeter axes (x or y). This was the orientation selected for the calculation and determination of the dependence of the degree of the luminescence polarization on the polarization of the exciting light. The degree of polarization was defined as $P = (I_1 - I_2)/(I_1 + I_2)$, where I_1 is the component parallel to the principal direction of the polarimeter, and I_2 is the component perpendicular to this direction.

Figure 20 shows the dependences of the degree of the luminescence polarization on the azimuthal angle η calculated for these orientations (the azimuthal angle gave the orientation of the radiation electric vector of the exciting light and it was measured from the horizontal). The orientation of the projections of the molecular oscillators on a plane perpendicular to the direction of excitation and observation of the luminescence is given alongside each curve. The

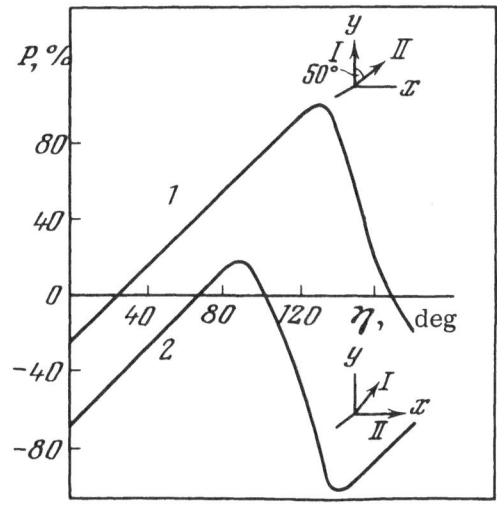

Fig. 20. Calculated dependences of the degree of polarization P of the luminescence on the orientation of the radiation electric vector of the exciting light: 1, 2) orientations of a crystal shown schematically alongside each curve.

plane on which these projections were made was selected to be perpendicular to one of the optic axes of the crystal. This selection had an important aim of avoiding difficulties associated with the birefringence of the exciting light. When this light was directed along an arbitrary axis in a crystal, the different velocities of the split components gave rise to different values of the net radiation electric vectors at different points along the propagation path. Clearly, this could "smear out" the required dependence irrespective of the presence or absence of energy migration. This masking effect could be avoided by the use of very thin crystals satisfying the condition $d\Delta n \ll \lambda$ (d is the thickness of the crystal, Δn is the birefringence, and λ is the wavelength of the exciting light). The thickness of such crystals should be of the order of several tens of microns, and they could be prepared only by sublimation. This was impossible in the case of naphthalene. Therefore, we used plates which were cut and ground from larger single crystals. In this way we were able to prepare plates a few tenths of a millimeter thick.

The undesirable effects of birefringence could also be avoided by directing the exciting light along the optic axis.

We used prepared spherical crystals from single-crystal blocks, and then we applied the Shubnikov (conoscopic figures) method to find the positions of the optic axes. Then, plates of minimum thickness, perpendicular to one of the optic axes, were cut and ground from these spheres. A specially constructed objective was used to direct a parallel beam of exciting light normally to the surface of the plate and along the optic axis, which avoided birefringence.

Information on the molecular orientation in the naphthalene lattice and also on the directions of the optic axes relative to the crystallographic axes of the crystal made it possible to calculate and to plot quite easily the projections of the luminescence oscillators of the impurity molecules onto a plane perpendicular to an optic axis. Such projections were included in the figures. The angle between the oscillator projections was 50°.

Naphthalene was transparent in the impurity absorption and luminescence regions, so that dichroism could not distort the polarization dependences. Moreover, naphthalene was not optically active. Therefore, it was a convenient host lattice for doped crystals. Clearly, the method described above was applicable only when the impurity molecules could be excited without any interaction with the host substance. Naphthalene activated with the selected phosphors was once again a very suitable substance from this point of view. Naphthalene was completely transparent to the radiation produced by a mercury lamp at 436 nm, which was absorbed strongly by the impurities.

Our measurements were carried out visually using a Cornu polarimeter. An important feature was the maximum possible reliability of the crossed position of the filters for the exciting light and luminescence. The exciting line at 436 nm was selected by two filters, one of which was a special filter for this line taken from a set supplied with the mercury lamp and the other was an interference filter. The green luminescence was measured after passing through a ZhZS-12 filter, which was selected from many tested filters and which gave a reliable crossed configuration with the filters for the exciting light. The phosphor impurities in naphthalene gave rise to high-efficiency green luminescence and they were thus suitable substances for the measurements described here.

The azimuthal dependence of the degree of polarization on the angle of rotation of a crystal about a horizontal axis coinciding with the optic axis under natural excitation conditions was needed for alignment purposes. The maximum positive degree of polarization of the luminescence corresponded to the coincidence of the bisector of the acute angle between the oscillators with the vertical axis of the polarimeter (configuration 3 in Fig. 21). Then, a crystal was rotated from this position by 25° (configuration 1) and by 65° (configuration 2). The

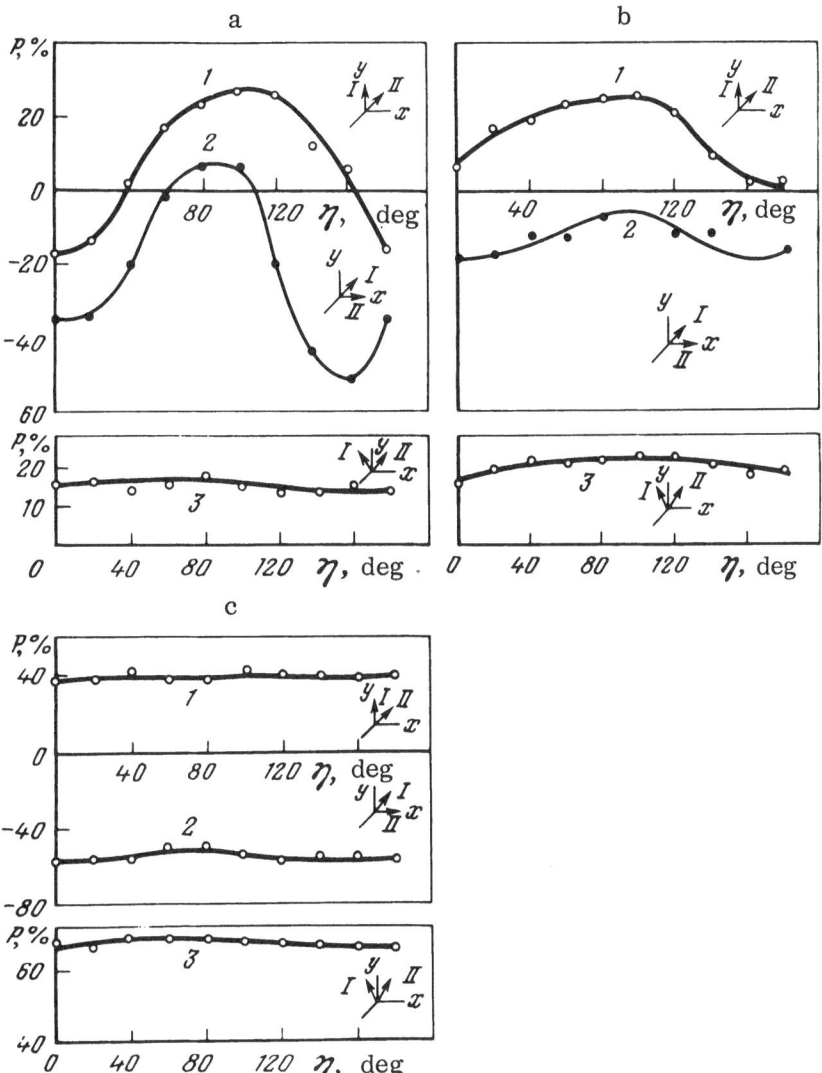

Fig. 21. Experimentally determined dependences of P on η for phosphor A in naphthalene: a) concentration 10^{-6} g/g; b) 8×10^{-6} g/g; c) 3×10^{-5} g/g. The orientation of a crystal is shown alongside curves 1, 2, and 3.

first of these positions corresponded to the orientation in which one of the luminescence oscillators coincided with the vertical axis of the polarimeter, whereas in the second position this oscillator coincided with the horizontal axis (corresponding to calculated curves 1 and 2 in Fig. 20).

The luminescence polarization was measured in these positions as a function of the angle η of the rotation of the Polaroid which passed the exciting light.

Since naphthalene easily sublimated in air, the high quality of the polished plate surfaces was maintained by making all these measurements in a special holder placed inside a fused quartz cell filled with distilled water. Moreover, this made it possible to eliminate or minimize the depolarization due to the scattering on the surface of a crystal.

The results of measurements carried out on naphthalene containing the phosphor A in concentrations of 10^{-6}, $8 \cdot 10^{-6}$, and $3 \cdot 10^{-5}$ g/g are plotted in Fig. 21 (curves 1 and 2). The impurity concentration was determined from the absorption spectra.

In these experiments it was extremely important to ensure that the polarized exciting light did not reach the polarimeter. Otherwise, the polarization of the exciting light would change considerably during the measurements and even a small contribution of the exciting light to the luminescence could give rise to a masking dependence. The most convincing control experiment was the measurement of the dependence of the luminescence polarization on the polarization of the exciting light in the orientation denoted by 3 in Fig. 21, i.e., when the bisector of the acute angle between the oscillators (corresponding to the maximum in the azimuthal dependence) coincided with the vertical axis of the polarimeter. We mentioned earlier that in this orientation of the crystal the expected dependence should not be observed even in the absence of energy migration. Consequently, in this case only the masking dependence due to the exciting light with modified polarization could be observed. The results of measurements carried out along these orientations were plotted in Fig. 21 (curves denoted by 3). They demonstrated that there was no significant contribution from the exciting light because the degree of polarization was not affected by the orientation of the radiation electric vector of the exciting light.

Let us now return to the main results. It is clear from Fig. 21 (configurations 1 and 2) that the strongest dependence of the luminescence polarization on the polarization of the exciting light was observed when the impurity concentration was 10^{-6} g/g, whereas for the $8 \cdot 10^{-6}$ g/g concentration it was weaker, and, finally, it was practically absent in the $3 \cdot 10^{-5}$ g/g case. The nature of the dependence was in satisfactory agreement with the theoretical calculations.

For comparison, we plotted in Fig. 22 the calculated curves for the orientations 1 and 2, normalized in amplitude; points and circles in Fig. 22 denote the experimental values. Clearly, the experimental dependences were in good agreement with the theory. Hence, we concluded that at concentrations of 10^{-6}-10^{-5} g/g, which corresponded to average intermolecular distances of (100-50)d (d is the lattice constant), there was no transfer of the excitation energy between the impurity molecules. A considerable migration occurred for shorter distances, which were of the order of (10-20)d, i.e., ~ 150 Å.

These calculated distances between the molecules were only the average values. Fluctuations in the distribution of impurity molecules were likely in a crystal. The formulas of classical statistical physics could be used to calculate quite readily the probability of such fluctuations. These calculations indicated, as in [19], that when the impurity concentration was 10^{-3} g/g the fraction of the molecules which were accidentally in neighboring lattice sites was 1% of the total number of impurity molecules, whereas when the impurity concentration was

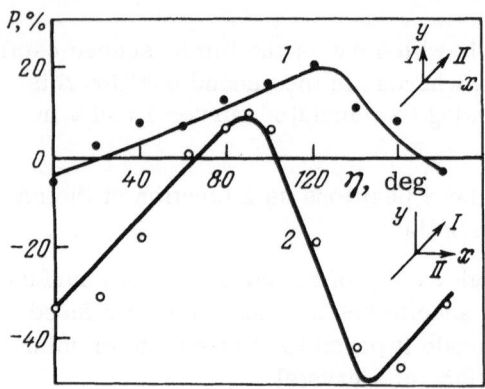

Fig. 22. Comparison of theoretical and experimental dependences.

$6 \cdot 10^{-5}$ g/g the fraction of such molecules was only 0.05%. These fluctuations could not affect significantly the observed polarization dependences.

A calculation carried out using the resonant transfer theory [20] allowing for the overlap of the spectra and for the mutual orientations of the molecules gave an effective distance of 50 Å between two phosphor molecules. Clearly, a comparison could be made with our experimental data only after averaging over all possible configurations. This could be done using formulas obtained in the theoretical investigation reported in [21].

In our case the donors and acceptors were identical molecules but they had different orientations. The total quantum efficiency of the energy transfer from a donor to an acceptor $D \rightarrow A$ calculated allowing for the averaging over R_{DA}, i.e., the probability of finding at least one acceptor molecule in the effective interaction region of a donor, is given by the formula

$$\rho = 1 - \exp[-n_A V_{eff}],$$

where $V_{eff} = (4/3)\pi R_{eff}^3$, $n_A = C_A/a^3$, C_A is the relative concentration of acceptors, and a^3 is the volume of a unit cell. If $R_{eff} \approx 50$ Å and $C_A \approx 10^{-4}$ g/g, we find that $\rho \approx 30\%$.

Calculations indicated that these parameters were sufficient to destroy the dependence $P_{lum}(P_{exc})$. Thus, we concluded that the migration of energy between impurity molecules occurred in accordance with the induction–resonance mechanism.

However, one should discuss the possibility of pairwise incorporation of impurity molecules in the host lattice. Several different direct and indirect experimental results testified against this tendency. These results were briefly as follows. The parameters of the luminescence emitted by an impurity in bulk differed very greatly from the luminescence emitted when the same substance was an impurity in a different host. This difference was independent of the presence of a second impurity in the host lattice. The second impurity did not affect the temperature dependence of the luminescence intensity. The ratio of the intensities of the luminescence emitted by two impurities (donor and acceptor) depended on the absolute concentration when the relative concentration was 1 : 1 (such a dependence would not be observed if the molecules were located side by side in pairs). The polarization diagrams of an impurity were independent of its concentration within a wide range (10^{-3}–10^{-7} g/g). The polarization diagrams of one impurity were independent of the presence of a second impurity. An analysis of the fine structure of the low-temperature spectra of doped naphthalene crystals [22] revealed the presence of bands not only of single molecules but also of pairs and triplets (clusters) but it also showed that the concentration of the latter was low and in full agreement with the random distribution model. When the concentration was reduced to 10^{-4} g/g, the cluster bands disappeared practically completely from the spectrum.

Finally, a direct experimental proof of a random distribution of single impurity molecules was provided by the polarization dependences described above.

Thus, a luminescence polarization study of energy migration led to a very important conclusion on the structure of doped crystals, which could not be obtained as reliably by other experimental methods.

The reduction in the amplitude of the dependence $P_{lum}(P_{exc})$ was partly due to unavoidable experimental errors, particularly because the direction of the exciting light could be oriented along the optic axis only to within 1–2°. Naphthalene exhibited a considerable birefringence [23], but its influence near the optic axis was slight. The directions corresponding to 1–2° with respect to the optic axis gave rise to a change Δn amounting to a few ten-thousandths. Estimates indicated that the resultant difference between the paths traveled by the components and the rotation of the net radiation electric vector of the exciting light could not suppress the calculated dependence $P_{lum}(P_{exc})$ but could reduce its amplitude somewhat. The necessary condition, derived in [4] for the thickness of the investigated crystals, $l < 2c/\Delta n\omega$, gave in our case an estimate $l < 2$ mm, which was satisfied in our conditions. Control experiments carried out on plates 2 mm thick yielded dependences which were smoothed out considerably.

All these results were confirmed by experiments carried out on naphthalene single crystals containing other activators (phosphors). The polarization dependence $P_{lum}(P_{exc})$ disappeared in approximately the same range of concentrations for all the phosphor activators.

The proposed luminescence polarization method is clearly the only method for detecting energy migration between identical impurity molecules incorporated in the host lattice of a molecular crystal.

The results obtained demonstrate also that polarization methods can be used to study not only the orientation but also a wider range of structural properties of doped molecular crystals and of physical processes occurring in these crystals.

Literature Cited

1. N. D. Zhevandrov, V. I. Gribkov, and V. K. Gorshkov, Izv. Akad. Nauk SSSR, Ser. Fiz., 32:1346 (1968).
2. N. D. Zhevandrov and T. V. Il'inykh, Zh. Prikl. Spektrosk., 15:1041 (1971).
3. N. D. Zhevandrov, Izv. Akad. Nauk SSSR, Ser. Fiz., 20:553 (1956).
4. N. D. Zhevandrov, Tr. Fiz. Inst. Akad. Nauk SSSR, 25:3 (1964).
5. V. N. Varfolomeeva and N. D. Zhevandrov, Opt. Spektrosk., 5:571 (1958).
6. G. S. Belikova, V. N. Varfolomeeva, and N. D. Zhevandrov, Izv. Akad. Nauk SSSR, Ser. Fiz., 29:1326 (1965).
7. G. S. Belikova, V. N. Varfolomeeva, and N. D. Zhevandrov, Kristallografiya, 13:129 (1968).
8. G. S. Belikova, V. N. Varfolomeeva, and N. D. Zhevandrov, Kristallografiya, 13:267 (1968).
9. G. S. Belikova, N. D. Zhevandrov, T. V. Il'inykh, L. E. Kraeva, and N. V. Khromeeva, Kristallografiya, 17:366 (1972).
10. A. I. Kitaigorodskii, Organic Crystal Chemistry [in Russian], Izd. Akad. Nauk SSSR, Moscow (1955).
11. N. M. Melankholin and S. V. Grum-Grzhimailo, Methods for Investigating Optical Properties of Crystals [in Russian], Izd. Akad. Nauk SSSR, Moscow (1954), p. 170.
12. G. S. Belikova and L. M. Belyaev, in: Growth of Crystals (ed. by A. V. Shubnikov and N. N. Sheftal'), Vol. 3, Consultants Bureau, New York (1962), p. 316.
13. E. Sackmann and D. Rehm, Chem Phys. Lett., 4:537 (1970).
14. R. Pariser, J. Chem. Phys., 24:250 (1956).
15. E. Sackmann, Chem. Phys. Lett., 3:253 (1969).
16. G. S. Belikova, N. D. Zhevandrov, T. V. Il'inykh, L. E. Kraeva, and N. P. Sul'zhenko, Kristallografiya, 18:779 (1973).
17. G. S. Belikova, N. D. Zhevandrov, E. V. Zharavova, T. V. Il'inykh, and L. E. Kraeva, Kristallografiya, 19:1204 (1974).
18. N. D. Zhevandrov and T. V. Il'inykh, Izv. Akad. Nauk SSSR, Ser. Fiz., 36:970 (1972).
19. V. I. Gribkov, Thesis for Candidate's Degree [in Russian], Lebedev Physics Institute, Academy of Sciences of the USSR, Moscow (1965).
20. M. D. Galanin, Zh. Eksp. Teor. Fiz., 28:485 (1955); Tr. Fiz. Inst. Akad. Nauk SSSR, 12:3 (1960).
21. S. I. Golubov and Yu. V. Konobeev, Fiz. Tverd. Tela, 13:3185 (1971).
22. V. L. Broude and A. V. Leiderman, ZhETF Pis'ma Red., 13:426 (1971).
23. S. Bhagavantam, Proc. R. Soc. A, 124:545 (1929).

INFLUENCE OF A STRONG ELECTROMAGNETIC FIELD ON PHASE TRANSITIONS IN FERROELECTRICS

B. P. Kirsanov

A report is given of a theoretical analysis of the influence of a strong but non-resonant optical field on phase transitions in ferroelectrics. It is shown that the nonlinear properties of a ferroelectric give rise to an additional ordering of elementary dipoles, and, as a result, the transition temperature rises. It is found that, in some cases, the phase transition apparently smears out. An estimate is obtained of the transition temperature rise as a function of the field intensity.

Introduction

Linear or nonlinear interaction of light and matter alters the free energy and other thermodynamic functions of matter. Optically induced ordering of dipoles may be equivalent to cooling, etc.

The action of light and the resultant changes in the thermodynamic properties may, in principle, give rise to phase transitions such as condensation of a vapor into a liquid, transformation of a metal into a dielectric, transitions to the superconducting state, crystallization, transitions from paramagnetic to ferromagnetic or diamagnetic state, transitions from paraelectric to ferroelectric state, transitions accompanied by a change in the symmetry, etc. We shall confine our attention to optically induced phase transitions in transparent ferroelectrics.

The effective field in an isotropic or cubic medium is [1]

$$\mathbf{E}_g = \mathbf{E} + \frac{4\pi}{3}\,\mathbf{P}, \tag{1}$$

where $\mathbf{E}$ is the external field. The polarization in the case of nonlinear interaction in a centrosymmetric medium satisfies the relationship

$$P = \alpha_1(E + fP) + \alpha_2(E + fP)^3 + \alpha_3(E + fP)^5, \tag{2}$$

where $f = 4\pi/3$; α_1 is the linear susceptibility, and α_2 and α_3 are the nonlinear susceptibilities.

A ferroelectric phase transition may be regarded as the appearance of a moment P in the absence of a field E, i.e., it can be regarded as self-polarization of matter giving rise to a spontaneous polarization P_s. When a ferroelectric is cooled below the Curie temperature T_c, a change in the parameters α_i and T due to the lowering of T results in a transition from the paraelectric state (phase) for which P_s is zero to the ferroelectric state with P_s.

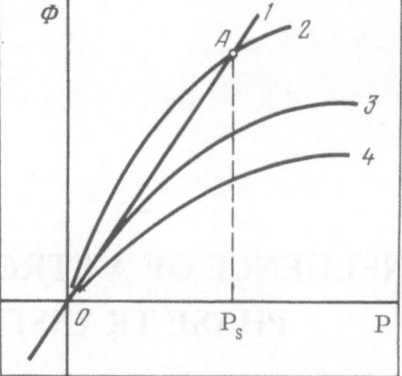

Fig. 1. Graphical determination of spontaneous polarization for a transition of the second kind: 1) Φ_1; 2) Φ_2 ($T < T_c$); 3) Φ_2 ($T = T_c$); 4) Φ_2 ($T > T_c$).

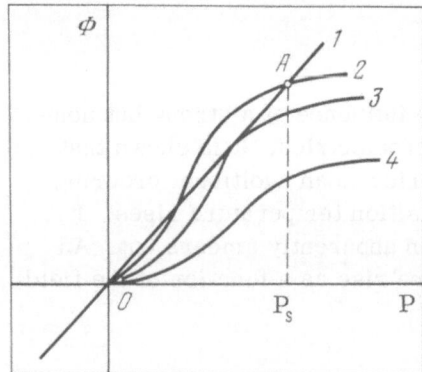

Fig. 2. Graphical determination of spontaneous polarization for a transition of the first kind: 1) Φ_1; 2) Φ_2 ($T < T_c$); 3) Φ_2 ($T = T_c$); 4) Φ_2 ($T > T_c$).

This phase transition may be of the first kind (BaTiO$_3$) or of the second kind (K$_2$H$_2$PO$_4$, usually denoted by KDP).

The ferroelectric transition can be explained readily by solving Eq. (2) graphically. This can be done by plotting the left- and right-hand sides of Eq. (2), denoted by Φ_1 and Φ_2. The solution of Eq. (2) for a transition of the second kind in $E = 0$ is plotted in Fig. 1. It is clear from this figure that, in the range $T > T_c$, the graph representing the right-hand side Φ_2 does not intersect the straight line $\Phi_1 = P$ because $f\alpha_1 < 1$. This means that the polarization P is not self-sustaining (Φ_2 is less than Φ_1), and the resultant spontaneous polarization is unstable and tends to zero. If $f\alpha_1 > 1$, Φ_1 intersects Φ_2 at a point A. Between 0 and A, we find that Φ_2 lies above Φ_1, and the resultant spontaneous polarization generates a stronger polarization, i.e., it becomes self-enhancing so that the system assumes the state with P$_s$ (point A).

A phase transition of the first kind is illustrated in Fig. 2. In this situation, the system passes through a barrier at temperatures $T_c < T$ and assumes a stable state with P$_s$ (point A).

Thus, the additional intersection of Φ_1 and Φ_2 in Fig. 2 corresponds to a phase transition.

More detailed information on ferroelectrics can be obtained from monographs [1-3]. We shall now consider possible mechanisms of optically induced ferroelectric transitions.

§ 1.   Optically Induced Ferroelectric Transitions

We shall first find the change in the critical transition temperature on application of a static or alternating field, and we shall use the linear approximation. We shall assume that the paraelectric and ferroelectric phases are in equilibrium, i.e., the pressures are equal and the thermodynamic potentials are the same for both phases:

$$\varphi_1(E, T) = \varphi_2(E, T).$$

Then,

$$\frac{\partial(\varphi_1 - \varphi_2)}{\partial T_c} dT_c + \frac{\partial(\varphi_1 - \varphi_2)}{\partial E} dE = 0.$$

Bearing in mind that the specific entropy S and the specific polarization P are

$$\frac{\partial\varphi}{\partial T} = -S, \quad \frac{\partial\varphi}{\partial E} = -P,$$

we obtain

$$-(S_1 - S_2) dT_c - (P_1 - P_2) dE = 0, \quad \frac{\partial T_c}{\partial E} = -\frac{P_2 - P_1}{S_2 - S_1}. \tag{3}$$

(The last equation is of the Clausius−Clapeyron type.) It should be noted that the entropy decreases on transition to the ferroelectric phase because of ordering and $-\Delta S_{21} = -(S_2 - S_1) > 0$. The change in the entropy $-\Delta S_{21}$ amounts to about 0.1-1 cal·mole^{-1}·deg^{-1} for most substances. If the field is optical, E is the amplitude of the electric vector of the radiation field $E(\omega)$, and the polarization is $P = \varkappa(\omega)E(\omega)$, where $\varkappa(\omega)$ is the linear susceptibility. Then, the shift of the transition temperature follows Eq. (3):

$$\Delta T = \left| E \frac{P_2 - P_1}{\Delta S_{21}} \right| = |E|^2 \left| \frac{\varkappa_2(\omega) - \varkappa_1(\omega)}{\Delta S_{21}} \right| = |E|^2 \frac{\Delta n_{21} n_0}{2\pi \Delta S_{21}}, \tag{4}$$

where n_0 and Δn_{21} are, respectively, the refractive index and the change in this index at the ferroelectric transition point.

Example. In the case of BaTiO$_3$, we have (in cgs esu) the parameters $n_0 = 2.4$, $\Delta n_{21} = 0.04$, $\overline{\Delta S_{21} = 1.3} \cdot 10^5$, $|E|^2 = 2 \cdot 10^5$ (50 MW/cm^2), and we find that Eq. (4) yields $\Delta T = 0.02°$K.

We shall now consider nonlinear optical effects. The optical rectification (detection) effect may occur in noncentrosymmetric substances. This effect gives rise to an additional static polarization in an optical field:

$$P(0) = \xi \left[\frac{\varepsilon(\omega) + 2}{3} \right]^2 \left[\frac{\varepsilon(0) + 2}{3} \right] |E(\omega)|^2, \tag{5}$$

where ξ is a third-rank tensor which we shall assume, for simplicity, to be a constant. Its value is 10^{-6}-10^{-9} cgs esu. The factor $[(\varepsilon(\omega) + 2)/3]^2 [(\varepsilon(\omega) + 2)/3]$ represents the correction allowing for the effective field [4, 5]. Since $[\varepsilon(0) + 2]/3 \approx (4\pi/3)\varkappa(0)$ $[\varkappa(0) \ll 1$ near the Curie point], this static polarization is equivalent to the action of a static field

$$E(0) = \frac{4\pi}{3} \xi \left(\frac{\varepsilon(\omega) + 2}{3} \right)^2 |E(\omega)|^2. \tag{6}$$

It follows from Eq. (3) that in a transition of the first kind the transition-temperature shift due to the static field is ($P_1 = 0$ in the paraelectric phase)

$$\Delta T = \frac{\partial T_c}{\partial E} dE = -\frac{P_2}{\Delta S_{21}} \frac{4\pi}{3} \xi \left[\frac{\varepsilon(\omega) + 2}{3} \right]^2 |E(\omega)|^2. \tag{7}$$

In a transition of the first kind, the appearance of a static external field at temperatures $T > T_c$ is analogous to a shift of the curve Φ_2 in Fig. 2 to the left, and a sudden appearance of an intersection of the curves in Fig. 2 corresponds to the phase transition. Although the author is not aware of any real ferroelectric which is noncentrosymmetric in the paraelectric phase and exhibits a transition of the first kind, such ferroelectrics may exist.

Example. In the case of BaTiO$_3$ [2, 3], whose parameters are (in cgs esu) P$_2$ = 7.8 $\cdot$ 10^4, ΔS_{21} = 1.3 $\cdot$ 10^5, $\varepsilon(\omega)$ = n$_0^2$ = 2.4^2, ξ = 10^{-6}, $|E(\omega)|^2$ = 2 $\cdot$ 10^5, we find from Eq. (7) that ΔT = 1.6°K.

In a phase transition of the second kind in the presence of the nonlinear optical rectification effect — for example, in KDP in which light travels in the necessary direction ($\xi \neq 0$) — the transition smears out, i.e., such properties as the polarization, specific heat, etc., vary continuously with temperature. In fact, displacement of the curve Φ_2 to the left in Fig. 1 in the range T > T$_c$ results in an intersection with Φ_1 even in infinitesimally weak fields, and the point of intersection shifts continuously with rising E. In the absence of the optical rectification effect in the paraelectric phase of a centrosymmetric material or one with zero components of the tensor ξ (for example, KDP with light traveling along the [100] axis), the transition-temperature shift may be due to a change in the static susceptibility under the influence of the strong optical field. The change in the susceptibility is [4, 5]

$$\Delta = \chi \left[\frac{\varepsilon(\omega) + 2}{3} \right]^2 |E(\omega)|^2, \qquad (8)$$

where χ is a fourth-rank tensor whose components should be within the range 10^{-11}–10^{-14} cgs esu [6].

An additional change in the susceptibility should alter (increase) the transition temperature because the additional susceptibility tends to enhance the self-polarization effect. The effective susceptibility in Eq. (2) is then $\alpha_1 + \Delta$. It is clear from Figs. 1 and 2 that the additional susceptibility $\Delta > 0$ causes counterclockwise rotation of the curve Φ_2 so that the curves Φ_1 and Φ_2 intersect in the range T > T$_c$. The temperature shift can be estimated by expanding $\alpha_1(T)$ near the transition temperature T$_c$:

$$\alpha_1(T) = \alpha_1(T_c) + \left(\frac{\partial \alpha_1}{\partial T} \right)_{T=T_c} (T - T_c). \qquad (9)$$

In the case of ferroelectrics [2, 3], we have

$$\left(\frac{\partial \alpha_1}{\partial T} \right)_{T=T_c} = - \frac{1}{f^2 C}, \qquad (10)$$

where C is known as the Curie constant. Comparing Eqs. (8)–(10), we find the effective temperature shift

$$\Delta T = \left(\frac{4\pi}{3} \right)^2 \chi \left(\frac{\varepsilon(\omega) + 2}{3} \right)^2 |E(\omega)|^2 C. \qquad (11)$$

Example. In the case of BaTiO$_3$, we have (in cgs esu) $\varepsilon(\omega)$ = n$_0^2$ = 2.4^2 and C = 10^4 so that, if $|E(\omega)|^2$ = 2 $\cdot$ 10^5 and χ = 10^{-12} (approximately), we find from Eq. (11) that ΔT = 0.2°K.

Electrostriction can also alter the ferroelectric transition temperature in a strong optical field: The change may be a rise or a fall of T$_c$. We can show that the pressure dependence of T$_c$ in the case of transitions of the first kind is governed by an expression similar to the Clausius–Clapeyron equation [3]:

$$\frac{\partial T_c}{\partial p} = \frac{\Delta V}{\Delta S_{21}}, \qquad (12)$$

where ΔV is the change in the volume at the transition point.

In the case of $BaTiO_3$, we have $\partial T_c / \partial p < 0$ and this pressure coefficient amounts to $-6 \cdot 10^{-3}$ deg/atm [2, 3]. In a phase transition of the second kind [3], we have

$$\frac{\partial T_c}{\partial p} = \frac{(\beta_2 - \beta_1)}{(c_{p2} - c_{p1})} \frac{T_c}{\rho} \tag{13}$$

(Ehrenfest relationship), where β, c_p, and ρ are the volume expansion coefficient, specific heat, and density, respectively. Experiments carried out on KDP type crystals show that

$$\frac{\partial T_c}{\partial p} \sim 10^{-3} \text{ deg/atm.}$$

The pressure produced by the optical field is

$$p = \frac{1}{4\pi} \left(\rho \frac{\partial \varepsilon}{\partial \rho} \right)_S |E(\omega)|^2. \tag{14}$$

We can easily see that in an optical field with a power density ~ 50 MW/cm^2 ($|E|^2 = 2 \cdot 10^5$ cgs esu), the pressure corresponding to $[\rho (\partial \varepsilon / \partial \rho)]_S \sim 1$ is about $2 \cdot 10^{-2}$ atm, and, consequently, the transition temperature changes only slightly as a result of striction in such fields:

$$\Delta T = \frac{\partial T_c}{\partial p} \Delta p \sim 10^{-4} \text{ °K.}$$

§2. Heating Effects in a Strong Optical Field

An optical field may heat matter and thus prevent a phase transition. We shall now consider the most important heating effects.

Electrocaloric Effect

The application of an optical field at $T = $ const and for $V = $ const produces a heat

$$Q = \frac{|E(\omega)|^2}{4\pi} T \left(\frac{\partial \varepsilon}{\partial T} \right)_p.$$

In the adiabatic case, this alters the temperature by

$$\Delta T = -\frac{Q}{c_p},$$

where c_p is the specific heat at constant pressure. Since $T(\partial \varepsilon (\omega)/\partial T)_p \sim \varepsilon(\omega)$, it follows that

$$\Delta T = \frac{|E(\omega)|^2}{4\pi} \frac{\varepsilon(\omega)}{c_p}. \tag{16}$$

Example. In the case of $BaTiO_3$ with $\varepsilon(\omega) = 5.8$ and $c_p = 0.13$ cal $\cdot$ g^{-1} $\cdot$ deg^{-1} the application of a field such that $|E|^2 = 2 \cdot 10^5$ cgs esu produces a temperature rise $\Delta T \sim 0.01$°K.

Change in Temperature Due to Adiabatic Transition

In an adiabatic transition induced by an optical field, the temperature may rise by [2, 3]

$$\Delta T = \frac{dQ}{c_p} = \frac{\Delta S_{21} T_c}{c_p}. \tag{17}$$

Example. In the case of BaTiO$_3$ with c$_p$ = 40 cal$\cdot$g$^{-1}\cdot$ deg^{-1}, we have dQ = 50 cal/mole and ΔT $\sim$ 1°K.

Heating Due to Absorption

This heating effect changes the temperature by

$$\Delta T = \frac{Wkl}{V\rho c_p},$$ (18)

where k is the absorption coefficient, l is the length of the sample, V is the volume, ρ is the density, and W is the total energy of the transmitted radiation.

Example. In the case of BaTiO$_3$ with c$_p$ = 0.13 cal$\cdot$g$^{-1}\cdot$deg^{-1}, we find that at T = T$_c$ for a sample with l = 1 cm, k = 0.01 cm^{-1}, V = 1 cm^3, ρ = 6 g/cm^3, and W = 1 J, we obtain ΔT $\lesssim$ 0.01°K.

Conclusion

The example of a ferroelectric in a strong optical field shows that phase transitions may be induced by this field even in a transparent medium. The above estimates indicate that the effects should be observable. In the case of resonance absorption of light, one would expect much stronger effects because of the larger values of the susceptibilities.

Literature Cited

1.   C. Kittel, Introduction to Solid State Physics, 2nd ed., Wiley, New York (1956).
2.   F. Jona and G. Shirane, Ferroelectric Crystals, Pergamon Press, Oxford (1962).
3.   W. Kanzig, "Ferroelectrics and antiferroelectrics," Solid State Phys., 4:1 (1957).
4.   S. A. Akhmanov and R. V. Khokhlov (eds.), Problems in Nonlinear Optics [Russian translation], Mir, Moscow (1966).
5.   N. Bloembergen, Nonlinear Optics, Benjamin, New York (1965).
6.   A. P. Veduta and B. P. Kirsanov, Zh. Eksp. Teor. Fiz., 54:1374 (1968).